10/91

WILDFLOWERS OF THE LLANO ESTACADO

WILDFLOWERS OF THE LLANO ESTACADO

Francis L. Rose
and
Russell W. Strandtmann

TAYLOR PUBLISHING COMPANY
DALLAS, TEXAS

Copyright © 1986, Francis L. Rose and
Russell W. Strandtmann

All rights reserved.

ISBN 0-9617102-0-9

No part of this publication may be
reproduced in any form or by any
means without permission in writing
from the publisher.

All photographs by the authors.

Printed by: Taylor Publishing Company
1500 W. Mockingbird Lane
Dallas, Texas 75235

Publication Consultant: Karen Blakeley Camp

Printed in the United States of America

This book is dedicated to our mothers.

Florence McNeill Rose
Frieda Schulz Strandtmann

Foreword

The Llano Estacado is a huge mesa that occupies approximately 37,000 square miles (22,000,000 acres) in northwest Texas and eastern New Mexico. It is larger than the combined states of Maryland, Massachusetts, New Jersey and Rhode Island, and it is larger than South Carolina! The area is devoid of native trees save for the tiny shin oak and it is the remnant of a vast grassland prairie that once extended northward into Canada. Often called the Southern High Plains, the Llano Estacado is separated from the rest of the Great Plains Province by the west-to-east cut of the Canadian River and the north-to-south cut of the Pecos River.

To the naked eye the Llano Estacado appears flat; yet, it tilts gently to the southeast at 8-10 feet per mile. There is no geological relief except for the 34,000 or so shallow depression called playas. In its ocular monotony, the Llano Estacado is the largest isolated, non-mountainous geological area in North America, and it stands as one of the earth's major geological features.

Preface

As zoologists we have reveled for many years in the progressive color changes that carpet the Llano Estacado from spring to fall. We are not plant specialists, however, we did observe a few and the culmination of these observations is the book before you. Our observations were assembled not because we had to or because we should have done so but because we wanted to do it! We met no challenges, we set no goals but we do hope that the book will fill a void for those so disposed as to be interested in wildflowers.

The photographs were taken with 35 mm cameras using a macrolens and natural light or with a 135 mm lens, bellows and flash. Low light and a brisk wind are the mistress of a flower photographer; out-of-focus flicks with no depth are his illegitimate children. We have catalogued over 1200 slides (keepers) and would like not to have to say how many we threw away.

Obviously not all flowers (species) could, or should be included in such a treatment. We have included those that, in our judgement, would most likely be encountered by the lay person. Some can be seen in your yard, others in the alley behind your house, others in vacant lots, along roadsides, in parks and, if pressed, we must admit that a few, just a few, mind you, we dangled before the intrepid stalker. Some flowers you might stand among but never see, others you will chisel from between the cracks in your sidewalk, and others you will assault with steely pry in your lawn. The plant that you so admire in a vacant lot becomes a loathsome, fetid creature whilst nestled among your rose plants. The delicate yellow flower of the goat head plant is a harbinger of doom for the bicyclest but a living delight for those engaged in the selling of bicycle innertubes! There is a place for all wildflowers: 'tis not they that are confused!

Acknowledgements

Many people were helpful to us in developing this book but by and large, we did it ourselves, our way! Neither of us claims to be a botanist, a point we cannot reinforce too staunchly. All funds and time were provided at personal expense. We have received neither encouragement nor disappointment from our employers but by the nature of our employment we were provided ample opportunity to fail or succeed professionally. To Donna Precure-Rose and Ruth Strandtmann we owe a weighty debt, one not likely to be repaid. We thank Drs. Chester Rowell, David Northington, Raymond Jackson and Laurie Robbins for the identification of various plants. Drs. Chester Rowell and Marshall Johnson read early drafts of the manuscript. Especial thanks are due to the Riley Miller family, the John and Jack Boren families, the John Lott family, the Martin Parks family, and to Dr. Gerald Wollom for providing us with access to land under their dominion. Without access to their land (which is not on the Llano Estacado) there would have been no place to lick our wounds, clear our minds and purge our spleens. Ruth Strandtmann made the difference between a manuscript and a book. Her generosity, support and determination to see the project through publication made the difference. We owe much to many in this vast land: perhaps this book will help.

Table of Contents

Foreword vii
Preface ix
Acknowledgements x
Historical Perspective 1
Soils 3
Weather 5
   Precipitation 5
   Temperature 5
Habitats 7
   Grasslands 7
   Pastures 7
   Playas 8
   Sandhills 8
   Exposed Caliche 9
   Entrenchment Canyons 9
   Ecologically Disturbed Areas 9
Seasonal Succession 11
What Is a Flower? 13
Classification 15
Key to the Families of Flowering Plants 17
Preparation and Storage of Plants 27
Wildflowers by Family 29
Vocabulary of Botany 81
Flowering Schedules 87
Index 93

Historical Perspective

During the Cretaceous Period (72 to 135 million years ago) the area to become the Llano Estacado was covered by a vast inland sea. The deposits of this prehistoric sea are composed of fossiliferous limestone and claystone that dip to the southeast at 7-8 feet per mile, a gradient closely approximating the surface slope of today. Remnants of ammonites, snails, bivalve mollusks, heart urchins, oysters and pectins abound in these deposits. Following the retreat of the seas, many of the Cretaceous deposits were stripped away and whereas they can be observed in the area east and south of Post and in the canyon complex around Lubbock, they are not present, for example, in Palo Duro Canyon.

There appears to be some controversy as to the age of the Cenozoic deposits that overlay the Cretaceous material on the Llano Estacado. The Cenozoic deposits are generally thought to be Pliocene in age but faunal assemblages might turn out to indicate that some are Miocene in age. At any rate, these deposits represent the erosional remnants derived from the eastern face of the Rocky Mountains that were swept southeast by large, now extinct, rivers. The Cenozoic deposits have been termed the Ogallala Group and are composed of two primary layers. The Couch Formation represents the initial stages of Ogallala deposition that occurred in the pre-existing channels carved into the underlying deposits. The Bridwell overlays the Couch where present and represents the final filling of channels and the deposition of a relatively smooth surface graded toward the southeast. The Bridwell is topped off with a 30-foot thick calcareous deposit of caliche called the caprock. During maximum extension of the Llano Estacado, the area extended southeast, probably to as far as a line drawn in an arc between present-day Wichita Falls and Abilene, Texas.

No sooner than the Llano Estacado reached its maximum dimensions than did environmental forces set about its destruction. The Red, Brazos and Colorado rivers inched north and west, forming tall cliffs resilient to erosion on the eastern front and carving intrenchment canyons, some of which nearly transect the Llano Estacado. Although the Pleistocene Epoch that followed the Pliocene had its on-again off-again moist periods, the face of the Llano Estacado probably did not change excessively. Streams meandered here and there in their quest for the sea but by and large, a grassland was in place.

The separation of the Llano Estacado from surface water sources from the north (Canadian River) and the west (Pecos River) eliminated a powerful erosional force. The Bridwell caprock caliche resisted erosion in general and thus a relatively high, dry plain with no rivers or streams was formed.

Pleistocene soils were, in general, overlain with wind-blown sands (probably derived from the Pecos River bed to the southwest) and from extinct rivers (sand hills north of Littlefield). Generations heaped upon generations of grasses and the high rate of surface water capture (98%) combined with the wind-blown sand to produce the fertile, farmable soil so characteristic of much of the area today.

Soils

A person walking on the Llano Estracado will encounter various types of soil surfaces. Over large areas to the south and west, exposed caliche is the rule. This caliche represents the top surface of the Bridwell Formation that was either not covered by Pleistocene or Recent deposits, or the deposits were blown elsewhere. In areas around Bledsoe, Seminole, Brownfield and in general toward the south, the top soil often is composed of undulating small dunes of shifting, wind-blown sand. When not moistened, the grains of the sand tend not to stick together but trail through your fingers like sand in an hourglass. Below this loessic soil there can be a firmer, more compact soil, the origin of which must be determined locally. In some areas the fine wind-blown sand has been swept into playas, leaving the more compact soils exposed.

The playa bottoms are variable but in general they have a top layer of darker soil with a high organic content due to countless generations of decayed vegetation. These soils contract drastically on drying, forming wide and deep cracks. Immediately under the playa topsoil one can usually (depending on the age and successional stage of the playa) find a grayish Randall clay, formed from the wash-in particles surrounding the playa.

The tongue of wind-blown sand that extends from the western area of the Llano Estacado near Muleshoe toward the east peters out near Cotton Center in Hale County. This area is composed of relatively large wind-blown dunes. The origin of the sand might be from the beds of extinct rivers.

As one moves toward the north the topsoil particles become smaller and the soils are more compact. Presumably the smaller sand particles toward the north reflect their wind-blown origin from the southwest.

The actual topsoil overlaying the caliche varies in thickness and caliche can be exposed in the fields at almost any place. In fact, the depth of the topsoil reflects the primary difference in the quality of the farming soils on the Llano Estacado when compared to more northern areas of the Great Plains. The depth of the farmable soil on the Llano Estracado is from 2-4 feet while in, for example, Iowa, the topsoil can be 10-20 feet.

Weather

The Llano Estacado is classified as semiarid. This does not mean that the climatic characteristics are intermediate between mesic and xeric but rather that the time and amount of precipitation are basically unpredictable. Even so, a precipitation pattern exists. Another factor in characterizing the weather on the Llano Estacado revolves around the sheer size of the area. And, as to be expected, the weather varies with latitude and longitude.

Precipitation

The amount of precipitation increases going from west to east. The mean number of inches of precipitation in the extreme western section is 12-13 whereas the eastern segment revels in its 20-21 inches. The winter months are the driest with May being the wettest month. There appears to be a general lessening in the amount of precipitation per month throughout the summer and fall, until November again is reached. The greatest amount of water that falls on the Llano Estacado does so from spring, summer and fall thunderstorms.

Snow fall is sporadic and averages 14-15 inches in the north to only 2-5 inches in the south.

Temperature

The mean temperature, of course, increases in a north to south direction, ranging from 57-58 F in the north to 63-64 F in the Midland area. The mean date of the last freeze ranges from March 28-April 1 in the south to April 23 in the Dimmit area. The mean date of the first freeze in the fall ranges from October 17 in Amarillo to November 6-10 in the extreme south.

All in all, the climate is relatively dry and the temperatures are mild. The area is, or should be, noted for its most pleasant evenings throughout the spring, summer and fall months.

The wind provides a force that must be dealt with by plants and human inhabitants as well. Dust-laden skies generate outstanding sunsets but the wind, the constant howling wind, that churns and swirls the loessic sand until the sky and firmament appear as one, constantly reinforces the image of impermanence.

Habitats

Being a portion of the Great Plains Physiographic Province, the Llano Estacado is a grassland included in the short-grass district of the Kansan Biotic Province of Texas. Tales are legion about the grasses of yore but alas, most of the Llano Estacado suffers heavily under the plow.

It would be accurate to state that if the Llano Estacado had had the diversity of habitats, say as does South Carolina (Coastal Plain, Piedmont, mountains, rivers, streams, marshes) that we would not have attempted to organize this book. Our saving grace resides with the sameness of the area; yet, one can delimit a series of general habitats that have some practical value. We are not implying that these habitats are discrete, for they merge perceptibly.

Grasslands

There is little native grassland left on the Llano Estacado. Because of the flat lay of the land with the crust of friable, fertile topsoil, most of the area is dominated by farming and ranching enterprises. Where the grasslands are in evidence, usually the topsoil is absent or it is very shallow or it is mixed with caliche. More often than not, cattle graze on the grasses. The few large blocks of grassland are in protected areas such as the Muleshoe National Wildlife Refuge near Morton.

Pastures

This habitat is highly modified and is best depicted as a mesquite-short grass association. Although generally overgrazed, plants with sharp sticky or distasteful characteristics thrive here; they are plants that cattle avoid when feeding if at all possible. Whereas mesquite is the dominant tree, various cacti, buffalo gourd and devil's claw also excel here.

Playas

The term playa means beach in Spanish but to a geologist it refers to a catchment basin that holds water but that has no entry or exit stream. On the Llano Estacado there is approximately one playa per square mile. For the discussion here we are not including the large salt lake basins such as Bull Lake and Illusion Lake, which are classified as playas but are nonetheless entities unto themselves. The playas provide us with two arrays of flowering plants; those that are aquatic and those that grow in moist conditions. The seeds of many plants are washed into the basins and left to germinate as the water recedes. The primary indicator plants here include arrowhead plant, knotweed with its beautiful pink inflourescences and the tiny-flowered *Aster subulatus*. The playa and pasture habitats often merge to provide an interesting continuum.

Sandhills

These are the wind-blown, undulating mounds of sand that extend from near Muleshoe eastward into Hale County. The area is widest in the west and extends eastward as an ever narrowing ridge. This is a distinct habitate and on driving north, one comes onto the hills suddenly and leaves them suddenly. Along the northern edge, the dunes fade quickly into open grassland with little or no top soil. The sandhills are unique to the area and the show of flowers found there can be spectacular. Shin oak, the soap berry tree, hackberry trees and elms dot the landscape. Sage, sumac and plum bushes are everywhere in evidence.

Water is held in the porous sand and in recent times subsurface water was just a few shovel blade depths away. Because of the irregularities of the land surface, little cultivation had taken place. However, with modern technology and tenacity of purpose, more and more of these beautiful hills are being degraded into sameness.

Exposed Caliche

The absence of a top soil exposes the underlying caliche. This can be along a canyon, a road cut, a ditch or in an open field. More often than not, the flat, exposed caliche areas are pastures and whereas they look barren from the road, many plants thrive here. Spring is heralded each year by the beautiful carpet of yellow bladder pod, the white mountain daisy and the yellow *Haplopappus*. The end of the season is festooned with the beautiful purplish-red inflorescences of gay feather and the yellow-flowered broomweed.

Entrenchment Canyons

The tributaries of the Red, Brazos and Colorado rivers have etched into the Llano Estacado and in some instances, cut nearly across its surface. To the east these tributaries have carved out deep canyons and arroyos of immense proportions but toward the west they become shallow, but respectable, cuts into the surface. The walls of the steep canyons and arroyos provide a vertical association of plants with those requiring more water toward the base to those more tolerant to desiccation near the top. Water flows in some canyons and deep pools are present. As these streams, which are not part of the Llano Estacado, meander along, the water nourishes a profusion of wildflowers along its shoulders. Palo Duro Canyon is a study itself in vertical and seasonal succession of wildflowers.

Ecologically Disturbed Areas

You might be wondering, "Can these two people be serious?" We realize that this designation is somewhat vague but nonetheless it is functional. A tremendous number of wildflowers will be found in areas such as along highways, section roads, alleys, parks and vacant lots. Wildflowers seem to flourish in such areas and for this reason we feel that they should be listed.

The roadsides can be staggering in their number and coloring of wildflowers; yet, the area just beyond the fence is pale. Much has been written as to why the roadsides excel as "gardens" and although we will not dwell here, we can simply point out that the roadsides receive a double dose of water as runoff from the road surface. As most of the roadsides are lower than the road bed, water tends to collect there.

Seasonal Succession

The flowering process is complex but the events involved are not random! Some plants flower in the spring, others in the summer, and others only in the fall. A few, such as the geranium (*Erodium*), can flower in deep winter. Some plants have a narrow time frame in which they bloom, others trickle out their blossoms throughout the warmer months.

The three external factors that affect plant growth, flowering and seed set are temperature (soil and ambient), water availability and day length.

Given the proper temperatures, soil, surface and stored water must be adequate to foster proper growth and flowering. Some plants, such as the lilies, rush pea and gourds have a bulb or tuber in which is stored water and nutrients. Thus, these plants are not necessarily dependent upon rain to get the juices flowing. The extreme of storage can be found in the buffalo gourd that can have a tuber several feet below the surface of the soil and that can weigh some 20-odd pounds. In sandy areas the tuber of the deadly nightshade (*Solanum*) can reside 5-10 feet below the surface of the shifting sands.

If temperature and water conditions are adequate, plants can be placed into one of three groups based on their flowering response to photoperiod.

1. Short-day plants flower when the length of day is below a critical value — spring or fall.
2. Long-day plants flower when the length of day exceeds a critical value — summer.
3. Neutral plants flower when other factors are adequate and are independent of day length.

On the Llano Estacado, as elsewhere, there is a definite seasonal cycle or succession of flowering. Some plants one would only expect to find in flower in the spring and to look for them in the fall would be in vain. We have attempted to pinpoint when one would be most likely to first observe flowering in the more dominant plants.

Some will have a narrow period of time in which they flower (two weeks), others will trail out their efforts. By and large, a given plant species will initiate flowering with a high proportion of individuals soon participating. The peak period is followed by a long decline in activity until flowering has virtually ceased. The degree of trailing out is dependent upon the plant species and the availability of water. In essence, it might be feasible to try and pinpoint when a plant species will begin blooming and when the peak period of activity is likely to occur, but one is treading on thin ice when trying to predict when all flowering will have ceased. At any rate, refer to the Flowering Schedule and you will gain some notion as to when a given plant species can be expected to be in flower.

Some plants have short life spans once the seeds germinate. They burst from the ground, flower, set seed and die. Following seed set, the vegetative function of the plant is finished. These plants are called *annuals*. Another tactic is for an individual plant to produce seeds over a number of years. Once the business of reproduction is over the plant can use its time fixing the sun's energy and storing nutrients to give it a start the following year. These plants are called *perennials*.

What Is a Flower?

Flowers are a plant's way of saying "hi." And, in fact, they are the reproductive structures of the *sporophyte* generation. The floral parts that rapidly gather our attention are modified leaves. The true flowering plants, the *angiosperms*, have numerous ways to insure that fertilization will occur; early in their history many developed a novel trick, pollination by insects. This mutualism has spanned some 130 million years and, due to stringent competition, has led to the vast array of colorful flowers that we perceive today.

Some angiosperm plants produce flowers with both male (pollen) and female (ovary) components. Some produce only one type of sexual flower; others produce both types but at different times; and, still others produce both types but they are mutually infertile.

One would be hard-pressed to find a wildflower that is anatomically ideal but with all things being considered, there is an underlying theme of structure. The floral parts are arranged around the axis of the stem. The *sepals* are leaf-like, generally green and make up the first whorl of the flower. All the sepals together (single or fused) are called the *calyx*. The *petals*, the white or colored units that nestle down into the sepals are the units that first grab our attention. The petals can be separate (a rose) or fused (a morning glory). The petals form the *corolla*, which along with the calyx constitute the *perianth*.

Within the perianth one can find the *carpels*, the organs that are associated with seed production. In some plants the carpels are fused together to form a single *pistil*. The enlargement of the base of the pistil is the *ovary* and contains the ovules that develop into seeds. The slender portion of the pistil above the ovary is called the *style* and terminates in the sticky, pollen capturing *stigma*.

The male portions of the flower are called the *stamens* and each consists of the delicate stalk-like *filament* tipped with the pollen-containing *anther* that is usually yellow.

Flowers can vary radically from this basic plan and the novice and professional should marvel at the intricate fusion of nature to produce the different types of blossoms. A sunflower, for example, is not a flower but an *inflorescence,* a colony of flowers. The usually yellowish or white peripheral units that appear to be petals are *ray* flowers; the innter core of the inflorescence is composed of the seed producing *disc* florets. Little wonder that the name "composite" is applied to this vast and complex group of plants.

Classification

The species designation is the only classification category that has a biological basis — mutual fertility. Yet, different species of plants can form viable hybrids. All higher categories of classification are abstractions; they are groupings of convenience but with implied hope that they signify relatedness. Each species has a "two-word" name that is latinized. Thus the annual white aster that grows so profusely in the playas is dubbed *Aster subulatus*. The first portion of the name is the *genus* and the second is the specific *epithet*. The initial letter in the species epithet is not capitalized.

Genera are grouped into families, which are grouped into classes, which are grouped into phyla (animals) or divisions (plants) which are grouped into kingdoms.

```
Kingdom = Plantae ... Animalia
 Phylum = Anthrophyta ... Vertebrata
  Class = Spermatophyta ... Mammalia
 Family = Asteraceae ... Homidae
  Genus = Aster ... Homo
Species = subulatus ... sapiens
```

The lay person for whom this book is written would do well to concentrate on two categories — the family and the genus. The families of plants are relatively stable, at least for those witnessed on the Llano Estacado. Also, once one familiarizes himself/herself with the families of plants, that information will be of value almost anyplace in the United States — or Europe or northern Mexico for that matter. Botanists have systems to file their plants logically and functionally. A couple of systems are used by the larger herbaria but they are somewhat similar and are based on the perceived relatedness of the plant families. So then, the families are basic to obtain a sound handle on the plants of our area and for this reason we have worked up and include here a general key to the primary families of plants that one might encounter on the Llano Estacado.

Key to the Families of Flowering Plants

Dichotomous keys are an integral part of identifying organisms. They are designed to eliminate as well as to confirm the presence of characters. Think of keys as though they were maps and directions informing you of the way out of a maze. Sure, you will eventually learn to use short cuts but to begin properly, start at the beginning of the key. First ascertain if the plant is a monocot or dicot. Then proceed through the key to the families. The identifying and eliminating characters are in couplets. If a plant does not have the characteristics mentioned in one couplet then it stands to reason that you must seek aid from the other couplet. Only practice will help.

Key

Leaves generally with parallel veins. Flower parts generally in 3's (but not in grasses). Leaf bases generally sheathing the stem.
....MONOCOTYLEDONEAE
Leaves net-veined. Flower parts in 4's or 5's or more numerous; rarely in 6's or 3's. Leaf bases not sheathing, or if so, it is by the united stipules.
.... DICOTYLEDONEAE

Monocotyledoneae

1. Plants aquatic or growing in marshy areas. Leaves saggitate (arrowhead shaped). Carpels numerous per flower and free from each other.
....*ALISMATACEAE*
1. Plants terrestrial, or at least not restricted to marshy areas. Leaves variable but not saggitate. Carpels solitary per flower or if several they are united.................2
2(1). Ovary Inferior...3

2. Ovary Superior 4
3(2). Stamens 3. leaves sharply keeled along the mid rib, partly clasping, and one leaf directly above another, as if astride.
.... IRIDACEAE
3. Stamens 6. leaves not sharply folded lengthwise, at least not the basal ones and leaves not in well marked rows. Plants diverse in habits.
.... AMARYLLIDACEAE
4(2). Sepals and petals quite dissimilar from each other. The three sepals free and herbaceous, the petals delicate, ephemeral. The third petal may be reduced or absent.
.... COMMELINACEAE
4. Sepals and petals similar, flowers radially symmetrical.
.... LILIACEAE

Dicotyledoneae

1. Flowers many on a single receptacle, in dense heads with one or more rows of bracts subtending the heads.
.... ASTERACEAE
1. Flowers not as above. If several or many from a common involucre, or in a head-like aggregation, then the individual flowers not on a common receptacle 2
2(1). Mature plants without leaves. The stems are succulent, photosynthetic and bear spines in clusters.
.... CACTACEAE
2. Not as above; not cactus-like 3
3(2). Flowers small, inconspicuous, green or greenish. In our species the leaves are small, cylindrical and succulent when young, becoming spiny with age. Our species is a weedy annual breaking at the ground when mature and rolling with the wind, scattering seeds. (Tumbleweed).
.... CHENOPODIACEAE
3. Not as above 4
4(3). Flowers lacking true petals. Calyx present or absent and when present is generally corolla-like 5

4. Flowers with both petals and sepals 11
5(4). Flowers one to several in a cup-like or bowl-shaped involucre 6
5. Flowers single or in various types of clusters. If an involucre is present it is leaf-like and subtends only a single flower 7
6(5). Flowers dioecious (unisexual). Involucres somewhat fleshy, cup-like, pistillate flowers without petals or sepals, with a 3-lobed ovary, each lobe with one seed. Style 3-branched, each branch generally divided. Staminate flowers with or without corolla-like sepals and may consist of only one stamen. Plants generally milky sap.
.... *EUPHORBIACEAE*
6. Flowers monoecious (plants bisexual). Involucres not fleshy, generally enlarging with age and becoming papery. Calyx corolla-like, bell or trumpet-shaped. The base of the calyx tube constricted above the ovary, making it appear as if the ovary is inferior.
.... *NYCTAGINACEAE*
7(5). Ovary Superior 8
7. Ovary Inferior 10
8(7). Herbaceous, perennial vines, climbing by means of twining petioles. Flowers dioecious. Stamens and pistils many.
.... *RANUNCULACEAE*
8. Not vines, erect or branching herbs. Flowers monoecious 9
9(8). Sepals 4 or 5, pink to red. Stamens 4 or 5. Fruit a small, red drupe.
.... *PHYTOLACCACEAE*
9. Sepals 2 to 6, petal-like, white to pink. Leaves with sheathing stipules. Fruits, 1-seeded brown or black achenes.
.... *POLYGONACEAE*
10(7). Flowers pink and white, regular, in clusters, fruit A 1-seeded nutlet.
.... *SANTALACEAE*
10. Flowers highly irregular, the calyx united into a u- or s-shaped tube and bizarrely colored. Fruit a many

seeded, six-sided capsule, stems trailing.
....ARISTOLOCHIACEAE
11(4). Petals separate..12
11. Petals united at the edges, at least basally. This generally easily determined..31
12(11). Ovary inferior, the petals and sepals arise from the top of the ovary..13
12. Ovary superior, the petals and sepals attached at the base of the ovary. This generally easily determined
....15
13(11). Sepals 3, soon falling. Petals 6, large, white, papery. Stamens many. Leaves, stems and capsular fruit prickly. Sap yellow-orange.
.... PAPAVERACEAE
13. Sepals 4 or 5, free or united. If plants prickly then not with yellow sap...14
14(13). Petals 5 or 10, stamens numerous. Leaves sticky with barbed or scabrous hairs.
....LOASACEAE
14. Petals 4 (Rarely 5), stamens 8, stigma with 4 linear lobes. Foliage neither sticky nor prickly.
....ONAGRACEAE
15(12). Shrub with spiny, rigid leaves. Wood yellow. Flowers yellow, fragrant, clustered in the leaf axils. Fruit an orange to red, pea-sized berry.
.... BERBERIDACEAE
15. Not as above..16
16(15). Leaves simple, the margins entire.......................17
16. Leaves compound, or if simple at least some leaves have margins variously serrate, dentate, lobed or dissected..20
17(16). Sepals 4 or 5. Petals 5 (or 3), very unequal, the 3 upper distinct and basally narrowed, the 2 lower small and greenish or lacking. Stamens 4.
.... KRAMERIACEAE
17. Not as above. The petals regular..............................18
18(17). Stems, at least the upper portion, 4-angled. Calyx tubular, with 4 to 7 lobes, the tube persisting around the developing ovary so that it may appear as if the ovary is inferior. Petals 4 to 6, attached to the upper rim of

the calyx. ....LYTHRACEAE
18. Not wholly as above..19
19(18). Sepals 2. Petals 4 or 5. Stamens 4 to many. Leaves alternate or occasionally nearly opposite, fleshy. ....PORTULACACEAE
19. Sepals 4 or 5. Petals 4 or 5. Stamens as many as the petals and alternate with them. Leaves alternate or opposite, narrow but not fleshy. ....LINACEAE
20(16). Plants erect, ill-smelling and clammy. Leaves alternate, trifoliate. Sepals 4, petals 4, stamens 8 to 20. Fruit a slender, clammy pod. ....CAPPARIDACEAE
20. Leaves and fruit not clammy.................................21
21(20). Leaves compound 25
21. Leaves simple, although they can be variously lobed or dissected..22
22(21). Sepals 2. Small, petals 4, bright yellow and highly irregular. One or both outer petals spurred, the inner ribbed. ....FUMARIACEAE
22. Sepals 4 or more. Petals 4 or more, regular..............23
23(22). Perianth parts 5 to 20, showy, generally regarded as all sepals. Stamens many. Carpels many, borne on a central cone. .... RANUNCULACEAE
23. Stamens 10 or less. Carpels 5 or fewer 24
24(23). Sepals 5, petals 5. Stamens 5 or 10, might not all be fertile. Leaves opposite, palmately lobed of pinnatifid. Fruit 5-celled with greatly elongated beak. .... GERANIACEAE
24. Sepals 4, early deciduous. Petals 4, regular, varying from small and inconspicuous to large and showy, white, yellow or lavender. Stamens 6. Leaves alternate. Fruit never with long beak. .... CRUCIFERAE
25(21). Leaves even-pinnate, opposite. Petals 5, regular. Flowers yellow. Either stiff shrubby plants with dark green leaves and the odor of creosote, or herbaceous

with trailing stems.
....ZYGOPHYLLACEAE
25. Leaves trifoliolate, palmate or odd-pinnate. If even-pinnate then not as above 26
26(25). Corolla regular...27
26. Corolla irregular 29
27(26). Sepals 4, deciduous. Petals 4. Stamens 6. Sap watery and pungent. Fruit a linear or globose silique.
.... CRUCIFERAE
27. Sepals 5, petals 5, stamens 5 or 10 28
28(27). Corolla perfectly regular, white to pink or rose-purple, never yellow. Plants low and herbaceous. Fruit 5-celled, with an elongated beak.
.... GERANACEAE
28. Flowers small and in dense heads, or larger and not in heads. Corolla of individual flowers may be slightly irregular. Leaves pinnately or bipinnately compound. Plants with erect or trailing stems.
.... FABACEAE
29(26). Sepals 2, petals 4. One or both outer petals spurred basally. Flowers bright yellow. Stamens 6.
....FUMARIACEAE
29. Sepals 5, petals 4 or 5. Stamens 5 or 10..................... 30
30(29). Sepals 5, corolla-like, the upper spurred. Petals 4, the two upper with spurs projecting into the spur of the sepal and the free portion resembling the head of a rabbit in profile. Leaves palmately divided (Delphinium).
.... RANUNCULACEAE
30. Sepals 5, petals 5 or sometimes seeming as 4 by union of the lower two to form the keel. Plants frequently with prickles. Flowers often yellow. Leaves palmate, pinnate or bipinnate.
.... FABACEAE
31(11). Corolla irregular, ovary superior...................... 32
31. Corolla regular or essentially so. Ovary superior or inferior 36
32(31). Herbs with branching and decumbent or trailing stems. Viscid pubescent and strongly scented. Corolla 2-lipped. Fruit an okra-sized pod terminating in a long, curved and hooked beak.
.... MARTYNIACEAE

32. Not as above 33
33(32). Leaves compound. Fruit a legume or indehiscent pod.
.... FABACEAE
33. Leaves simple, margins entire or variously lobed or dissected 34
34(33). Stems square, leaves opposite, upper leaves generally without petioles. Flowers strongly 2-lipped, arising from the leaf axils. Stamens 2 or 4. Ovary 4-lobed, forming 4 nutlets.
.... LAMIACEAE
34. Not as above 35
35(34). Sepals 5, unequal, the 2 inner sepals usually larger and petal-like. Petals 3; the central one boat-shaped, the 2 lateral ones more or less wing-like.
....POLYGALACEAE
35. Sepals 2 or 5, free or united. Petals 4 or 5. Corolla generally strongly 2-lipped. If nearly regular then plants tall and coarsely hairy and woolly.
....SCROPHULARIACEAE
36(31). Ovary inferior 37
36. Ovary superior 39
37(36). Stems trailing or climbing, with tendrils. Flowers unisexual. Corolla regular. Leaves alternate, simple or compound, petiolate.
....CUCURBITACEAE
37. Plants without tendrils 38
38(37). Leaves opposite or whorled, simple. Flowers small, petals 4, white to pinkish. Fruit a capsule, about 2 mm long with few seeds.
....RUBIACEAE
38. Leaves alternate, simple. Sepals 5, large (10 mm or more long), showy, irregular. Fruit a many seeded capsule more than 4 mm long.
.... CAMPANULACEAE
39(36). Erect to prostrate herbaceous plants. Stamens many, united by their filaments into a column around the style. Style generally several-branched.
.... MALVACEAE
39. Stamens fewer (10 or less) and not united in a column around the style, at least not by their filaments 40

40(39). Stems generally square and erect or prostrate. Leaves opposite, without stipules, simple, entire or variously lobed, dentate or incised. Flowers small and solitary in the leaf axils or in densely clustered racemes. The corollas are rotate, 4 or 5 lobed with the lobes slightly unequal.
.... VERBENACEAE
40. Not totally as above 41
41(40). Low herbs with sour sap due to oxalic acid. Leaves trifoliolate, each leaflet notched at the apex. Flowers yellow or lavender, in loose umbels.
.... OXALIDACEAE
41. Not wholly as above 42
42(41). Plants with twining stems. Leaves alternate and long petioled. Flowers trumpet-shaped. In the parasitic dodder the stems are orange-yellow and leafless.
.... CONVOLVULACEAE
42. Not as above. If stems are prostrate, then not twining 43
43(42). Plants with milky sap. Leaves mostly opposite, without petioles or the petioles very short. Calyx-lobes rotate, strongly reflexed in some species. Corolla-lobes rotate or reflexed, somewhat fleshy. Fruit an okra-shaped follicle containing many flattened seeds, each with a tuft of silky hairs.
.... ASCLEPIADACEAE
43. Not with all the above characters 44
44(43). Plants herbaceous, erect, usually bristly. Leaves alternate, simple, entire, linear to slenderly ovate, sessile or with a short petiole. Corolla tube cylindrical, with spreading lobes. Stamens 5 and alternate with the lobes.
.... BORAGINACEAE
44. Not entirely as above 45
45(44). Erect herbs. Sepals 5, thin. Membranous, at least along the edges. Ovary 3-celled, the styles 3-cleft. Fruit a 3-celled capsule.
.... POLEMONIACEAE
45. Ovary 1, 2, or 4-celled. Style simple or at most with 2 stigmas 46

46(45). Erect, sometimes somewhat woody, herbs. Each flower with a single style and stigma. Fruit a berry or capsule. If a capsule then more than ¼ inch in diameter. Stamens 5, the anthers long, prominent and in some species joined around the style. Calyx in some species enlarges with age and completely surrounds the mature fruit.
....SOLANACEAE

46. Erect herbs. Each flower with 1 or 2 styles but always with at least 2 stigmas. Fruit a small 1- or 2-celled capsule about ¼ inch in diameter. Inflorescence in scorpioid cymes (Phacelia) or single in the leaf axils (Nama).
....HYDROPHYLLACEAE

Where appropriate we have used the species epithet for completeness but we urge no one to become aroused by such designations. Functionally, the lay person and botanist will most often refer to the genus unless there is a multitude of species involved. Common names have long been in vogue but such names for many plants on the Llano Estacado are meaningless — they provide no logical handle to grab onto! In some cases more than one unrelated plant has the same common name, tumbleweed, for example. In other instances the names were given to the plants in other geographic areas where there might be some association that is lacking here. Undoubtedly some of the names are European in origin and we cannot document that anyone from the Clovis culture to modern times ever used the name on the Llano Estacado, whether the plant was native or introduced. Where appropriate we suffered through the designations, modifying some and renaming others. When possible and plausible we were inclined to use common names of hispanic origin.

To us there appears to be no logical scheme in the structuring of common names. Some names might be hyphenated (tumble-weed), others are not. For simplicity we have eliminated the hyphen either whenever possible or plausible or when there just does not appear to be any structural reason to retain the hyphen!

Storage and Preparation of Plants

Often when you are out and about you will wish to collect plants for later identification. Be sure to grab more than just one blossom as the stem and arrangement of the leaves and even the root habit might aid your identification. For temporary collecting, the plants can be placed in a sealable plastic bag of appropriate size into which you have flicked a small amount of water from your fingers. DO NOT have free water in the bag, all that is required is enough to "fog" the inside. Inflate the bag, seal it and keep it cool if possible. Many plants and their flowers will keep for days in this manner if kept refrigerated. Some, such as blue flax, will not keep and the blossoms seem just to disappear. In warm weather a paper bag is preferred, as it does not accumulate as much heat from the sun.

For a more permanent record you will need to press and dry the plants. To do this, place the plant(s) along with a segment of the stem and root on a piece of newspaper. Spread the parts so they do not touch. You might wish to open and spread some of the flowers with a thin blade: a razor blade will do. Place the following information with the plant:

Name of collector:

Date that the collection was made:

Locality where the collection was made (include county, direction and distance from town, road number):

Remarks:

Do not rely on your memory. Too many valuable plants have gone to waste because of a lack of the aforementioned information. **When in doubt, Write it out!**

After folding the newspaper over on the plant you might wish to sandwich it between two pieces of cardboard and place a weight on top. For those more serious, a plant press is in order. Presses are simple to make but anything that will flatten and subsequently dry the plants will do. Properly pressed plants with the appropriate information might turn out to be of immense value to those maintaining herbaria.

   Although not for botanical use you might wish to try the following. Select the flower that you wish to preserve and place it in the following mixture in a can or jar (an oatmeal box works well). Use equal parts borax and cornmeal and add 3 tablespoons of salt per quart of mixture. Make sure that the petals are separated as you wish them to be when dried. Sprinkle the cornstarch mixture until the flower is covered and let it dry for three weeks. We have seen wildflowers dried in this fashion that were absolutely beautiful. Give it a try!

ARROWHEAD

WIDOW'S TEARS

SPIDERWORT

WILD ONION

CROW POISON

SPANISH BAYONET

RAIN LILY

BLUE-EYED GRASS

DUTCHMAN'S PIPE

WILD BUCKWHEAT

KNOTWEED

RUSSIAN THISTLE

SAND VERBENA

FOUR-O'-CLOCK

ANGEL'S TRUMPET

HIERBA DE LA HORMIGA

PIGEON BERRY

PURSLANE

FLAME FLOWER

PRAIRIE LARKSPUR

AGARITA

PRICKLY POPPY

SCRAMBLED EGGS

TANSEY MUSTARD

SPECTACLE POD

BLADDER POD

WALLFLOWER

CLAMMY PLANT

CAT CLAW

WHITEBALL ACACIA

SENSITIVE BRIAR

RUSH PEA

WHITE LOCO

SWEET CLOVER

WHITE CLOVER

FEATHER DALEA

JAME'S DALEA

GOLDEN DALEA

PRAIRIE CLOVER

LOCO WEED

CRAZY WEED

ALFALFA

RHATANY

PIN CLOVER

WOOD SORREL

YELLOW FLAX

BLUE FLAX

GOAT HEAD

MILKWORT

SNOW-ON-THE-MOUNTAIN

BULL NETTLE

WINE CUP

COMMON MALLOW

GLOBE MALLOW

CHEESE WEED

STICK-LEAF

PLAINS PRICKLY PEAR

TASAJILLO

CHOLLA

DEVIL'S HEAD

HEDGEHOG CACTUS

LOOSESTRIFE

AMMANNIA

EVENING PRIMROSE

EVENING PRIMROSE

GAURA

LIZARD TAIL GAURA

GREEN MILKWEED

ENGELMANN'S MILKWEED

BINDVINE

DODDER

BLUE GILIA

BLUE CURLS

TINY 'TUNIA

APACHE TEA

HELIOTROPE

PRAIRIE VERBENA

VERBENA

FROG FRUIT

HENBIT

LEMON BEEBALM

GERMANDER

GROUND CHERRY

PRAIRIE LANTERN

SILVER-LEAF NIGHTSHADE

WHITE NIGHTSHADE

BUFFALO BUR

FALSE NIGHTSHADE

JIMSON WEED

BEARD-TONGUE

INDIAN PAINTBRUSH

DEVIL'S CLAW

STAR VIOLETS

STAR VIOLETS

BUFFALO GOURD

PIEMELON

BALSAM GOURD

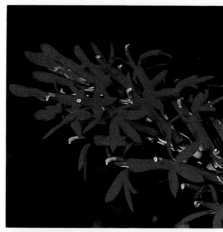
CARDINAL FLOWER

IRON WEED

GAY FEATHER

FALSE BONESET

CAMPHOR WEED

SLEEPY DAISY

CURLY-CUP GUM WEED

SAW-LEAF DAISY

BROOMWEED

HAPLOPAPPUS

JIMMY WEED

TANSEY ASTER

ANNUAL ASTER

MOUNTAIN DAISY

GREEN EYES

CUT-LEAF DAISY

RAGWEED

ZINNIA

MEXICAN HAT

PRAIRIE SUNFLOWER

BLUE WEED

COWPEN DAISY

COREOPSIS

THELESPERMA

RAYLESS THELESPERMA

FIREWHEEL

OLD RED EYE

BITTERWEED

SNEEZEWEED

DOG WEED

CLUSTER FLAVERIA

SAND PALAFOX

PAPER FLOWER

OLD PLAINSMAN

YARROW

THREADLEAF GROUNDSEL

STAR THISTLE

TEXAS THISTLE

DESERT HOLLY

SKELETON PLANT

SKELETON PLANT

GOAT'S BEARD

WILD LETTUCE

DANDELION

Wildflowers by Family

The wildflowers are arranged in the order of least specialized to most specialized. The number preceding the family name is the number allotted to that family by Correll and Johnston (1970; Manual of the Vascular Plants of Texas). When the family is represented by a single species or the species' characteristics are those of the family, the family treatment is not included.

The Monocots

23 ALISMATACEAE: WATER PLANTAIN FAMILY

Marsh plants with fibrous or tuberous roots. Leaves basal, with long petioles and linear or sagittate blades. Flowers in whorls on a central stem. Sepals 3, green. Petals 3, colored, deciduous. The tuberous roots are edible. Plantain is derived from plantar, the sole of the foot. The reference is to the shape of some of the leaves. The family name is derived from Gk., *alisma*, water plantain.

Sagittaria longiloba
ARROWHEAD PLANT   Color page 1, No. 1
The arrowhead or duck potato plant can be found in the playas and in roadside ditches if water is sufficient. The plants might be in dense stands and the creamy-white petals and arrowhead-shaped leaves protruding above the water surface are most often seen in summer-fall. Waterfowl feed on the tubers giving rise to the name of duck potato. The generic name is derived from Lat., *sagitta*, an arrow.

35 COMMELINACEAE: SPIDERWORT FAMILY

Perennials with tuber-like roots. Leaves alternate, flat. Stems nodose. Sepals 3, foliaceous. Petals 3 or 2, equal or unequal, colored, ephemeral. Named for J. and G. Commelin, Dutch botanists. There were three Commelin brothers, two became well-known botanists (= the two well developed petals), and the third was not so distinguished (= the weak petal of *Commelina*). Spiderwort refers to the appearance of the angular leaf arrangement that resembles a spider.

Commelina
WIDOW'S TEARS Color page 1, No. 2
The widow's tears or day flowers appear to have two dark blue, equal petals; however, close examination reveals a third, tiny white petal. The name reflects the fact that the beautiful flowers open for just one morning! They are found in moist, shaded areas and are common escapees from gardens. Flowering extends from late May into September.

Tradescantia
SPIDERWORT Color page 1, No. 3
The spiderworts can be found throughout most of the Llano Estacado, their blue 3-petaled flowers opening in early morning and closing in early afternoon. The plants do well when cultivated and appear to require considerable sun and water. Blooms from mid April through June. Most common in the sandhills where it occurs as clumps a foot or so tall. The genus is named for John Tradescant, a French traveler and gardener.

38 LILIACEAE: LILY FAMILY

Herbs with bulbs, corms, or tubers. Flowers in umbels or terminal racemes. The common garden *Asparagus* belongs to this family.

Allium
WILD ONION Color page 1, No. 4
The wild onions are hardy and are often considered pest plants. They appear in open areas on the prairie, in lawns and most places between. The umbels can have few to many flowers, which are pink to white. The plants extend upward about 6-8 inches and flower from March to July. If in doubt, tear it out — give it a sniff. The generic name is derived from Lat., *allium*, garlic.

Nothoscordum bivalve
CROW POISON Color page 1, No. 5
Quite similar to wild onion but without the odor. The whitish-yellowish flowers are usually fewer, larger and more loosely arranged than in the onion. Found in flat areas, pastures, as well as in towns. Flowers from mid-April through May. The generic name is derived from Gk., *nothos*, bastard, spurious, and *skordion*, a plant smelling like garlic. Apparently the name reflects someone's dismay at the lack of a garlic or an onion odor. Onion and garlic have been used by many cultures for culinary and medicinal purposes. Although we have not verified its use since the mid-1940's, garlic was used to ward off evil spirits, especially vampires. A small clove was placed in a hollow, perforated, tear-shaped locket and hung around one's neck. Some garlic was hung in windows for the same purpose. Poultices of garlic/onion, with a little camphor for "penetration," along with a generous helping of Black Draught laxative was a sure remedy for colds and kept at bay the dreaded consumption and "swamp fever." Unfortunately, sulfur compounds found in these plants make their way from the digestive to the circulatory system, where they are expelled in the breath and perspiration. Sir John Harrington wrote long ago in "The Englishman's Doctor":

> Garlic then have power to save from death
> Bear with it though it maketh unsavory breath,
> And scorn not garlic like some that think
> It only maketh men wink and drink and stink. (1609)

Yucca angustifolia
SPANISH BAYONET Color page 1, No. 6
The large (6-8 feet) *Y. treculeana* is not found naturally on the Llano Estacado but there are plants to be seen south and southeast of Post. Several species of the smaller yuccas occur here but the primary one is probably *Y. angustifolia*. Its 3-4 foot-high stalk and creamy white flowers are seen in May and early June. It grows well in compact soils with lots of caliche.

39 AMARYLLIDACEAE: AMARYLLIS FAMILY

Leaves fleshy, strap-shaped in *Cooperia*. Flowers single or racemose, borne on long stems. Sepals and petals similar. Fruit a 3-celled capsule with many black, disc-shaped seeds. Ovary is inferior. The name amaryllis is derived from that of the shepherdess in Virgil's Eclogues.

Cooperia drummondii
RAINLILY   Color page 2, No. 1
This plant flowers from May to fall after heavy rains. With their spectacular cream colored petals the 10-inch high plants are found primarily in pastures or upper shoulders of playas. The bulb is deep below the surface (12 inches). In general, each blossom opens near nightfall and withers before noon. Note that it is not a true lily. Named for Cooper, head gardener at Wentworth House, Yorkshire.

41 IRIDACEAE: IRIS FAMILY

Leaves grass-like (in blue-eyed grass) with flowers of 6 "tepals" (the sepals and petals similar). Ovary inferior, fruit a 3-celled, 3-lobed capsule. Iris was the goddess of the rainbow and messenger of the gods.

Sisyrinchium
BLUE-EYED GRASS   Color page 2, No. 2
A most beautiful blue flower with a bright yellow or gold throat. The flowers are smaller than a dime and occur at the tips of the 10-inch-high grass-like tufts. It occurs in patches associated with hard-packed soil loaded with caliche. It flowers from the first of April through May. The generic name is derived from Gk., *sys*, a pig, and *rhynchos*, a snout. Pigs must have been observed feeding around its roots.

The Dicots

57 SANTALACEAE: SANDALWOOD FAMILY

A perennial from a stoloniferous rootstock. Parasitic. Leaves linear, entire, small. Flowers white and pink, in racemes or corymbs. A widespread family found in Eurasia and in North America. The sandalwood tree belongs to this family but does not occur in North America. It has a compact, close-grained, yellowish, fragrant heartwood and has been much exploited for ornamental carvings and cabinet work. The family name refers to sandal.

Comandra pallida
BASTARD TOADFLAX (Not pictured)
Look for this small plant on the slopes of the entrenchment canyons. We have found it only once and include it here as it is a sole representative in our area of an interesting plant family. It is a perennial herb from a stoloniferous root stock. Leaves small, linear and entire. Flowers without petals, the sepals petaloid, small, pink and white, in corymbs. It blooms near midday but is "fickle," depending on cloud cover. It probably is not rare but rather its scarcity is due to the fact that no one looks for it. The generic name is derived from Gk., *come*, hair, and *andros*, a male, and apparently refers to the hairy attachment of the anthers to the sepals.

58 ARISTOLOCHIACEAE: BIRTHWORT FAMILY

Stems long, slender, prostrate, arising from deep, succulent roots. Leaves long-petioled, ovate to cordate, somewhat fleshy. Petals lacking, sepals petaloid and highly irregular, brownish-purple with yellowish spots and blotches.

Aristolochia coryi
DUTCHMAN'S PIPE Color page 2, No. 3
You will have to look close for this plant and its most spectacular flower. It is found near the base of entrenchment canyons in cool, moist places among large boulders.

It blooms in June. The name is derived from Gk., *aristos*, best, and *locheria*, parturition.

60 POLYGONACEAE: KNOTWEED FAMILY

Leaves alternate, elongated, with sheathing stipules. Calyx of 2 to 6 separate, often colored sepals. Petals lacking. Fruit a 3-angled or lens-shaped achene. Inflorescence in terminal or axillary racemes. The name is derived from Gk., *polys*, many, and *gonu*, a knee or joint, and apparently refers to the numerous and conspicuous nodes on the stems.

Eriogonum
WILD BUCKWHEAT　　　　　　　　Color page 2, No. 4
A tall (3 feet), somewhat spindly plant, is common along unmowed roadsides and railroads, and flowers from early July through September. Wild buckwheat reaches its greatest densities in the sandy soils but also occurs on caliche. A low-growing (6 inches) species is found on the caliche slopes and flowers in late summer and fall. The generic name is derived from Gk., *eryon*, wool, and *gonu*, a knee or joint, referring to the downy joints of the stems.

Polygonum bicornis
KNOTWEED　　　　　　　　　　　Color page 2, No. 5
A common plant found primarily in playas and in roadside ditches. The stems are erect but spindly, branched and the stem nodes are swollen, giving rise to the common name. The stems are dark red. It is an annual and can literally fill a playa! Flowering time is acutely tuned to precipitation but if conditions are right the dazzling display of pink that fills the playas is a sight to behold. Basically a summer bloomer. In the winter the tall plants (up to 5 feet) collapse, forming an intertwining mat.

61 CHENOPODIACEAE: GOOSEFOOT FAMILY

Flowers inconspicuous, usually green. Stems succulent when young, leaves generally small and entire. The name is derived from Gk., *chenos*, a goose, and *podes*, foot, and refers to the shape of the leaves.

Salsola kali
RUSSIAN THISTLE, TUMBLEWEED Color page 2, No. 6
A conspicuous floral component of the Llano Estacado. The roundish plants can be 6 feet in diameter and, unless you have driven through the area in the early winter with the wind howling, dodging these huge plant "balls" you have missed a vital component of living in the area. European in origin. *Salsola* is the diminutive of the Lat., *salso*, salty.

63 NYCTAGINACEAE: FOUR-O'CLOCK FAMILY

Leaves opposite, petiolate, without stipules. Flowers without true petals. The calyx is trumpet-shaped and colored, corolla-like. The flowers of many species open late in the day (thus the name) and the calyx withers and falls the next morning. The family includes the cultivated four-o'clocks and *Bougainvillea*. The generic name is derived from Gk., *nyx, nyktos*, night.

Abronia
SAND VERBENA Color page 3, No. 1
Found primarily in the moist, sandy areas. It is common in the sandhills and along the banks of streams in the canyons. The white to pink flowers occur in heads and open in late afternoon. Blooms primarily in April through June but occasionally trails into the fall. The generic name is derived from Gk., *abros*, delicate.

Mirabilis albida
FOUR-O'CLOCK Color page 3, No. 2
These plants can be found most frequently in sandy areas, along road cuts or on the canyon rims. The dull pinkish flowers are inconspicuous, opening in early evening and closing in early morning. The plants can attain a height of 4 feet but usually they are about 20 inches tall. The generic name is derived from Lat., *mirabilis*, wonderful.

Acleisanthes longiflora
ANGEL'S TRUMPET Color page 3, No. 3
The plant is inconspicuous and has trailing branched stems low to the ground. However, the flowers are spectacular. Each creamy-white flower is solitary and has an exceptionally long tube (thus the name). It flowers at night and the plants are nearly impossible to find during the day. Found along canyon rims generally in pastures around agarita or juniper. It doesn't seem to be on caliche. Blooms from late May to July. Probably serviced by moths. Rare. The generic name is derived from Gk., *akleistos*, not closed, and *anthos*, flower.

Allionia incarnata
HIERBA DE LA HORMIGA (ANT) Color page 3, No. 4
This plant was found by us below the canyon rims. It is low and trailing, almost vine-like. The purplish flowers are striking and slightly larger than a nickel. It seems to open just before noon and blooms from June through September. The genus is named for Carlo Allioni, an Italian botanist.

65 PHYTOLACCACEAE: POKEWEED FAMILY

Leaves alternate, entire. Perennial rootstock. Petals lacking, the sepals 4 (in our species) and colored. Flowers in axillary or terminal racemes. Fruit a berry, orange or red in color. The family name is derived from Gk., *phyto*, a plant or plants, and *lac*, lacquer.

Rivina humilis
PIGEON BERRY Color page 3, No. 5
The stems are generally less than two feet tall. Common to shaded areas in towns and along shaded stream beds in the canyons. The genus is named for A.Q. Rivinus, a German botanist. Actually his name was A. Bachmann; "Bach" in German means a little creek or rivulet, so he Latinized it to "Rivinis," which means creek in Latin, for botanical purposes. All else considered, he was serious about his earthly endeavors!

67 PORTULACACEAE: PURSLANE FAMILY

Leaves fleshy or succulent. Inflorescence of 2 sepals and 5 or more petals. Fruit a capsule of few to many seeds. Purslane is probably derived from porcelain and probably refers to the bright and shiny quality of the petals in some species.

Portulaca mundula
PURSLANE, MOSS ROSE　　　　　Color page 3, No. 6
An annual plant that emerges from a taproot. The alternate leaves are succulent. Noticeable white hairs emerge from the leaf axils. The purple-red petaled flowers are terminal and close to the ground. The growth habit is prostrate or ascending, sometimes as a single stem. Flowers throughout the spring and summer, depending on water availability. Found on the canyon slopes and in pastures. The name *portulaca* is Latin for little gate, referring to the lid on the seed capsule. Or, it is derived from Lat., *porto*, I carry, and *lac*, milk.

Talinum lineare
FLAME FLOWER　　　　　Color page 4, No. 1
This plant is most often found near the canyon rims nestled in the shade of an agarita or juniper. The flower is bright orange and the stem bears thin elongated leaves. The plant is less than 10 inches in height and the flower opens late in the afternoon. A tuber about the size of one's thumb rests about 5-6 inches below the surface. It is edible if one does not mind the taste of dirt! The generic name is derived from Gk., *thalia*, a green branch.

72 RANUNCULACEAE: CROWFOOT FAMILY

Leaves lobed or divided, basal, alternate, or opposite. Stamens many. Sepals 4 to many, frequently petal-like. Petals absent in *Clematis* but present and irregular in *Delphinium*. The young stems and leaves and the mature seeds of *Delphinium* are toxic to animals. The generic name is the diminutive of *rana*, Gk. for frog, referring to the marshy places often inhabited.

Delphinium virescens
PRAIRIE LARKSPUR Color page 4, No. 2
Occurs in pastures and along roadsides. The distinctive flowers are white and the plant grows to a height of about 10-15 inches. The upper sepal is prominently spurred. The generic name is derived from Gk., *delphin*, and refers to the resemblance of the spur to a dolphin's head. The common name refers to the similarity of the flower to the spurs of a lark.

Clematis drummondii
OLD MAN'S BEARD (Not pictured)
We have not observed this plant on the Llano Estacado. In the canyons it is a common vine covering fences and fence posts. The silky 2-4 inch long styles are white and form a fuzzy fruiting head that is visible in the late summer and fall. The generic name is derived from Gk., *klema*, a vine branch.

73 BERBERIDACEAE: BARBERRY FAMILY

Berberis trifoliolata
AGARITA, ALGARITA Color page 4, No. 3
Low shrub with alternate, trifoliate, stiff and spiny-margined leaves. Flowers are yellow, sweet smelling, in compact, axillary racemes. Fruit a tart, red berry that makes a passable wine and an excellent jelly. The trick is to separate the berries from the sharp spiny leaves! To do this, place a blanket under and around the bush. Beat the bush with a stick and save the booty in a bucket. After you have finished thrashing the bushes, spread the blanket out and dump the berries in the center. Wait for a strong wind (it will not be long) and then throw the berries and leaves into the air by snapping the blanket. The wind should blow the leaves away leaving the more dense berries to collect on the blanket. *DON'T DO THIS IN YOUR YARD!* The generic name is Arabic, *berberys*.

78 PAPAVERACEAE: POPPY FAMILY

Argemone sclerosa
PRICKLY POPPY Color page 4, No. 4
Leaves alternate, without petioles, frequently clasping.

Sap yellow-orange. Flowers large and showy, sepals 3, petals 6, stamens many. Fruit a large, many-seeded prickly capsule. The white, large, lacy petals are set off by the bright yellow of the stamenal complex. Blooms from late April through early September. Most often seen in pastures and open fields. Attains a height of about 2-3 feet. Not common on the Llano Estacado. The family name is from Lat., *papaver*, which means poppy. The generic name is derived from Gk., *argema*, cataract. Apparently it was used as a folk medicine to treat cataracts.

79 FUMARIACEAE: FUMITORY FAMILY

Corydalis curvisliqua
SCRAMBLED EGGS Color page 4, No. 5
Leaves alternate, dissected. Flowers in cymes, sepals 2, small and bract-like. Petals 4, in pairs, the outer pair spurred or saccate at base. Most often observed in sandy areas in shade. Although not rare, it is inconspicuous and is often found in less traveled areas. Fumitory is derived from Latin and means smoke of the earth. The flowers of our plant look like scrambled eggs — don't strain too hard on this one! The generic name is derived from Gk., *korydalos*, a lark. Apparently the spur of the flower looks like the crest of a lark.

80 CRUCIFERAE: MUSTARD FAMILY

Sap generally pungent. Leaves alternate, simple to variously divided. Sepals and petals 4 each. Stamens 6, unequal (4 one size, 2 another). Fruit a usually dehiscent silique. Called crucifer because the 4 petals form a cross!

Sisymbrium
TANSEY MUSTARD Color page 4, No. 6
Along with *Descurainia*, it is among the first plants to send up their stalks. It blooms from mid-January and by June covers yards and alleys. The tiny flowers are intricately beautiful but are often overlooked. Initially the vegetative growth is a rosette, resembling a large dandelion. The seed pods below the flowers appear to be tiny beans. Tansey is derived from Gk., *thanatos*, an instinctu-

al desire for death. One muses that our tanseys must share visually similar characters with the true tansey.

Dithyrea wislizenii
SPECTACLE POD Color page 5, No. 1
A plant associated with sandy areas. It reaches a height of about 3 feet and stands as a single stalk with a terminal flower head. The flowers are whitish and the bilobed seed pods, which look like spectacles, are distinctive. Blooms late April through September. The generic name is derived from Gk., *dys* twice and *thyreoeides* (a large oblong shield + *eidos*, form) and probably refers to the two shield-like seed capsules.

Lesquerella
BLADDER POD Color page 5, No. 2
One of the most common early bloomers. It virtually covers the caliche-rich areas. Plants are about 6-8 inches in height and the flowers are yellow. Blooming begins in early March, reaching a peak in April and trailing into May. Its numerous round seed pods are distinctive. The genus is named for Leo Lesquereux, a Swiss-American botanist.

Erysimum capitatum
WALL FLOWER Color page 5, No. 3
This is a rare plant on the Llano Estacado but it is spectacular when present. We have seen it in the sand hills along the road from Sudan to Nickels gin. The plant attains a height of 2-½ feet and probably blossoms from April through May into June. The generic name is derived from Gk., eryo, to draw. Apparently the sap of some species produces blisters.

Capsella bursa-pastoris
SHEPHERD'S PURSE (Not pictured)
A common inhabitant of alleys and vacant lots. It is uncommon out from towns. Its tiny white flowers give way to small heart-shaped seed pods. If the seed pods are opened after maturing they yield the bright yellow seeds — the shepherd's gold! Look for it in January and February. The generic name is the diminutive of the Lat., *capsa*, box. The "bursa-pastoris" alludes to a shape like a ram's scrotum, used as a purse by poor shepherds who

could not afford to use the "good" parts of the sheep's skin.

81 CAPPARIDACEAE: CAPER FAMILY

Polanisia
CLAMMY PLANT   Color page 5, No. 4
Typically with an ill smelling odor, palmately compound leaves and glandular-viscid pubescence. Flowers 4-merous. It grows in disturbed areas rich in caliche to the sandy bottoms of canyons. It is a spectacular plant when in bloom in late summer and fall. It reaches a height of about 1-½ feet. Pull some of the plant and handle it — it feels damp, soft, sticky, and peculiarly cool = clammy! The origin of the family name is from Gk., *kapparis*, but its meaning has been lost. The generic name is derived from Lat., *Poly*, name of a plant, and the Gk., *anisos*, unequal.

91 FABACEAE (LEGUMINOSAE): LEGUME FAMILY

Trees, shrubs, vines or herbs! Fruit a pod, either indehiscent or dehiscent along both sutures. A large family, with many species being poisonous to livestock. Beans, of one type or another, form a dietary staple for human beings throughout much of the world. Legume means pod. *Faba* is Latin for bean. Having evolved the trick of fixing nitrogen, the legumes inhabit some of the most inhospitable areas of the earth's surface.

Acacia
CATCLAW   Color page 5, No. 5
Shrub with much branching, usually less than 5 feet in height. It is armed with recurved "cat claws" that can tear the toughest fabric. The flowers form creamy white "balls" about the size of a marble. Generally, catclaw has somewhat elongated flower heads; not as round or "ball-like" as in *Acacia texensis*. The generic name is derived from Gk., *akazo*, I sharpen.

Acacia texensis
WHITE BALL ACACIA   Color page 5, No. 6
This plant is uncommon, grows low to the ground and does not have spines. Usually it is found in prairie grassland, sometimes on caliche and its growth habit is vine-like, with the runners emerging from a single stem. Possibly confused with *Schrankia* but its leaves do not close when touched. Flowers in late summer. *Acacia seyal* is probably the Hebrew shittah tree from which the ark and fittings of the Hebrew Tabernacle were thought to have been made.

Schrankia uncinata
SENSITIVE BRIER   Color page 6, No. 1
Generally a low running vine with pink flower "balls." The plant is prickled with numerous recurved spines. The leaves close quickly when touched. It blooms from early May to mid-August. Usually is is found in open fields with considerable caliche. The genus is named for Franz von Paula von Schrank, a German botanist.

Hoffmanseggia glauca
RUSH PEA   Color page 6, No. 2
A small plant (6 inches) with yellow flowers with a splash of brown. It is common along roadsides and in compact soils with or without high caliche content. It blooms from mid-May to early September. The genus is named for Johann Centurius von Hoffmansegg, a German botanist.

Sophora nuttalliana
WHITE LOCO   Color page 6, No. 3
A small (6 inches) plant inhabiting roadsides. It is found in clumps throughout the Llano Estacado. Its white blossoms remind one of hyacinths from a distance. It blooms from the end of April through May. Considered poisonous to livestock. The generic name is Arabic for leguminous tree.

Melilotus
SWEET CLOVERS   Color page 6, No. 4
The sweet clovers are common roadside plants. They are introduced but well established. The plants grow to about 2 feet in height. *Melilotus officinalis* has yellow flowers and blooms from late April into September.

Melilotus albus has white flowers and blooms somewhat later than *M. officinalis*. The generic name is derived from Gk., *meli*, honey, and *lotus*, lotus.

Trifolium repens
WHITE CLOVER   Color page 6, No. 5
This perennial forms large mats and the leaves are palmately trifoliate. Flower heads are about as wide as long and stand erect on weak stems. The corollas are white with a tinge of pink in most. Found primarily in municipal areas where it thrives in lawns and in parks. Sometimes called shamrock (which it isn't), a four-leaved representative has been known to impart as much good luck as a rabbit's left rear foot. This needs to be researched in greater depth, however, especially from the jaundiced eye of the rabbit! Trifolium refers to the three leaves.

Dalea frutescens
FEATHER DALEA   Color page 6, No. 6
A small shrub usually less than 3 feet in height. The flowers are small and purple and the plant occurs in open pastures and on caliche outcrops. It blooms from early April to August. The daleas are a conglomerate of many species with diverse growth habits. They are named for an English botanist, Samuel Dale.

Dalea jamesii   Color page 7, No. 1
This is a low-growing vine-like plant inhabiting caliche areas along canyon rims. Its yellow flowers are found in May. Rare.

Dalea aurea
GOLDEN DALEA   Color page 7, No. 2
A low to tall plant occurring along canyon stream beds and along road cuts. Its yellow flower heads are seen in June. The crushed plant has the odor of lemon.

Petalostemum multiflorum
PRAIRIE CLOVER   Color page 7, No. 3
This plant can have single or multiple stems emerging from a common base. The flowers are white and it blooms from early June into July. It is found primarily on exposed caliche and along sandy stream beds in the canyons. The generic name is derived from Lat., *petalo*, leaf, and Gk., *stemon*, warp, in the sense of stamen.

Astragalus
LOCO WEEDS Color page 7, No. 4
There are several species of *Astragalus* residing on the Llano Estacado. *Astragalus mollissimus*, the primary one, grows as a rosette in pastures and along roadsides. It does well in exposed caliche. The deep purple flowers are observed from March into June. The generic name is derived from Gk., *astrom*, a star, and *galu*, milk. The plants do not have a milky sap and it is thought that the name refers to the belief that some of these plants growing in a pasture will cause increased milk production.

Oxytropis lambertii
LOCO WEED, CRAZY WEED Color page 7, No. 5
A perennial herb that is usually solitary but can be in small colonies. Sporadic in occurrence and uncommon on the Llano Estacado, it is usually observed along roadsides or along road cuts in the caprock. The 10-25 flowers per raceme extend upward considerably more than do those of *Astragalus*. The pink-purple to lavender flowers appear in late May and June and are conspicuous. This loco weed is thought to be among the most deadly of the group, accounting for considerable livestock losses due to its being palatable and its wide distribution. The generic name is derived from Gk., *oxys*, sharp, and *tropis*, a keel. The keeled petal is sharply pointed.

Medicago sativa
ALFALFA Color page 7, No. 6
This is a common bushy green plant found primarily along roadways. It is about 2 feet tall and has purple flowers. It blooms from mid-May into September. The generic name is derived from Gk., *medike*, from Medea, the country of supposed origin.

92 KRAMERIACEAE: RHATANY FAMILY

Krameria lanceolata
RHATANY, HEEL BUR Color page 8, No. 1
A low shrub, intricately branched and trailing. Leaves small, alternate. Flowers in axils. Sepals 4 or 5, petals 5, very unequal. Fruit a 1-seeded pod. A common plant most often found scattered around in the mesquite —

short grass pasture and in juniper — agarita associations. The dark red flowers appear in early May into July. Named for J.G.H. Kramer, a German botanist. Rhatany refers to the dried root of two American shrubs, *Krameria triandra* and *Krameria argentea*, used as an astringent.

93 GERANIACEAE: GERANIUM FAMILY

Erodium cicutarium
PIN CLOVER Color page 8, No. 2
Leaves opposite, lobed, stipulate, long petioled. Flowers 5-merous. Styles persistent, developing into a long beak on the 5-lobed ovary. On drying, the twisting seed pod is spiraled into the ground. This introduced plant (Mediterranean area) is among the earliest bloomers. Its deep green, lacy rosettes are seen throughout the winter months and the tiny pink flowers can appear in January, although March and April appear to be the peak blooming times. It occurs in yards, alleys, roadsides and fields. In moist areas it tends to be more robust. The stork's bill, *E. texanum*, which occurs on some of the outlying mesas is native but we have not observed it on the Llano Estacado. Geranium means crane. The generic name is derived from Gk., *erodios*, heron.

94 OXALIDACEAE: WOOD SORREL FAMILY

Oxalis
WOOD SORREL Color page 8, No. 3
Leaves ternately compound. The three leaflets fold at dusk, during cloudy or hot weather. Can be used in salads with discretion. This tiny, yellow-flowered plant blooms from mid-February to mid-June. When the ripe pods are touched, they literally explode, strewing tiny round seeds upward to a foot away. Sorrel refers to sour, reflecting the high content of oxalic acid in the plant.

95 LINACEAE: FLAX FAMILY

Leaves simple, linear, with short or no petioles. Flowers 5-merous, in cymes. Seeds flat, oily. Two members of the group are found here. Flax, of course, refers to the fibers of some species that are used for spinning and weaving fabric.

Linum rigidum
YELLOW FLAX   Color page 8, No. 4
Yellow flax has flowers about the size of a nickel, which are displayed along spindly stalks. The petals can be nearly copper colored and have several dark lines converging toward the center. It blooms from late April to September and is usually found in flat, open areas — pastures, along roads in tall grass.

Linum pratense
BLUE FLAX   Color page 8, No. 5
A plant with small, light-blue blossoms about the diameter of a pencil eraser. It is found in flat, tall grass areas generally in more shade than the yellow flax. This plant can be quite tall but it is spindly and tends to lie down. The tiny round seed pods on the blue flax are distinctive. The flowers drop from the stems soon after collected.

96 ZYGOPHYLLACEAE: CALTROP FAMILY

Tribulus terrestris
GOAT HEAD   Color page 8, No. 6
Leaves stipulate, opposite, even-pinnate. Flowers 4- or 5-merous. Fruit of goat head with strong spines, capable of puncturing bicycle tires. It's name means "troubler of the earth" and it is a common invader of disturbed flat areas. It inhabits lawns, old fields and the silly thing thrives in sidewalk cracks! No soil is too hard or too devoid of nourishment to thwart this plant. In fact, it probably could be killed with kindness! The small yellow flowers are seen from the end of May through September. Caltrops are four-spined devices that, when thrown or placed on a flat surface, will always have one spine projected upward. They were used to persuade horses, and vehicles with pneumatic tires, to go elsewhere. The family name is derived from Gk., *zygo*, pair, and *phyllon*, leaf, an allusion to the opposite leaflet.

101 POLYGALACEAE: MILKWORT FAMILY

Polygala alba
MILKWORT Color page 9, No. 1
Leaves simple, entire, sessile or with short petioles. Flowers in terminal or axillary racemes. Sepals 5, unequal. Petals 3, united at base. Fruit a 2-celled capsule compressed contrary to the partition. Seeds pubescent. This is a low-growing plant (up to 12 inches) with white flowers. It blooms from mid-April to mid-June and occurs on compact soils and exposed caliche. The generic name is derived from Gk., *polys*, many, and *galu*, milk, and apparently refers to the plant's rumored value in increasing milk production in humans and livestock.

102 EUPHORBIACEAE: SPURGE FAMILY

A large and diverse family. Its flowers are unisexual. Sap milky in *Euphorbia*. Seeds of the castor bean plant, which is widely used as an ornamental, resemble engorged ticks and are poisonous. Spurge refers to purge. *Ricinus communis* produces the pale, viscous castor oil so admired by children as a cathartic or purgative.

Euphorbia marginata
SNOW ON THE MOUNTAIN Color page 9, No. 2
Found primarily in sandy areas, especially in the sand hills and along railroads and roadsides in sandy areas. The tall (up to 3 feet) plants have a single stem that branches above, and it flowers from late July through the fall. The flowers are tiny but the showy green bracts, trimmed in white, are eye-catching. This type of euphorb is commonly cultivated. CAREFUL — MAY CAUSE SKIN IRRITATION IN SOME PEOPLE. Euphorbus was a physician to the king of Mauretania.

Cnidoscolus texanus
BULL NETTLE, MALA MUJER Color page 9, No. 3
Not observed by us on the Llano Estacado, however, others have told us of its presence. We suspect that they have confused it with *Solanum rostratum*, the buffalo bur. If it does occur here it would be found in sandy areas and around pastures. The ensuing rash after contact with this

plant is aggravating to say the least. It would bloom in June and the flowers would be creamy white and smell like oranges. The generic name is derived from Gk., *knide*, nettle, and *skolos*, thorn.

115 MALVACEAE: MALLOW FAMILY

Herbs with simple, alternate and stipulate leaves. Petals 5. Stamens many, united into a column surrounding the pistil. Cotton, okra and hibiscus are common mallows. Mallow is derived from middle English, *malwe*, from Gk., *malakos*, soft, referring to its demulcent properties or to its soft, downy leaves.

Callirhoë involucrata
WINE CUP, COWBOY ROSE   Color page 9, No. 4
This spectacular mallow is deep red to nearly purple with a bright yellow stamenal complex. It is found in low, flat, moist grassland. Along canyon bottoms it can attain heights of two feet if supported by tall grass. In the sand hills it can be locally common but its flowering schedule is closely attuned to water availability. We have never observed it around playas but have seen it in moist areas associated with the entrenchment canyon drainages. The flowers open about mid-day and close near dusk. Flowers from mid-April into July. The generic name is derived from Gk., *kallirrhoe*, beautiful-flowing. Kallirrhoe was the wife of Alcmaeon, and caused his death because she coveted the famed necklace of Harmonia.

Malva neglecta
COMMON MALLOW   Color page 9, No. 5
This is a common plant on the Llano Estacado but it is restricted to urban areas. The plant has short, prostrate stems branched from the base. The leaves have a long petiole, are orbicular to reniform in shape, shallowly lobed and crenate. The plant might form a low, flat mat about 6 to 12 inches in diameter. The dark green leaves, flat growth habit and distinctive leaves make this plant easy to identify when it is not in bloom. The five petals are whitish, with four or so deep lavender stripes. The blossoms are rarely seen because they open near midday. Often seen in parks where it tolerates full sun light.

Sphaeralcea
GLOBE MALLOWS, COPPER MALLOWS
Color page 9, No. 6
These copper-colored flowers are exceptionally beautiful when viewed closely. The shorter (about 12 inches high) form begins flowering in April and usually is through by late May. It is a common plant along roadsides, alleys and fields. A taller type (2 feet) begins flowering a little later around mid-May and continues through June. It appears to be less common than the short form and is found in nearly the same habitats but with a greater penchant for caliche. The generic name is derived from Gk., *sphaira*, a globe, and *alcea*, mallow.

Sida lepidota
CHEESE WEED
Color page 10, No. 1
This plant is common but the flowers are rarely observed. It inhabits playas and literally rings the lower moist areas. The white-yellowish, with a touch of pink, flowers open around noon. The plant has low growth (6-8 inches) and blooms from early June into September. The generic name is derived from Gk., *side*, a kind of plant.

130 LOASACEAE: STICK-LEAF FAMILY

Mentzelia
STICK-LEAF
Color page 10, No. 2
Leaves ovate to oblong, both surfaces with barbed hairs. Fruit dry, capsular, with many seeds. This aptly-named plant is a true indicator of exposed caliche. It excels along roadside cuts and opens only in late afternoon to early evening. Two stick-leafs occur in the area, a smaller (2 feet) plant with slightly more of an orange tinge to the petals. The larger plant extends upward to 4 feet and its large cream-colored flowers are indeed beautiful. The plants bloom from May through September but on any given year, the smaller stick-leaf initiates the process. The genus is named for G. Mentzel, a German botanist. The family name is a Latinized version of a South American Indian word.

131 CACTACEAE: CACTUS FAMILY

Sap mucilaginous, stems thick, pulpy, photosynthetic, and with many spines. Leaves small and evanescent. Ovaries inferior, sepals, stamens, and petals many. Fruit (tuna) pulpy and edible. The young cladophylls and mature tunas are sold in some markets. Wine and jelly are made from the ripe tunas. The name cactus was applied by the Greek philosopher, Theophrastus, but members of the original genus have all been shifted to other genera. The name is derived from Gk., *kaktos*, but of course the ancient Greeks were unaware of what we now call cactuses, since the plants are from the New World.

Opuntia macrorhiza
PLAINS PRICKLY PEAR   Color page 10, No. 3
A typical, low growing, flat cladophyll cactus. The bright yellow flowers appear in early May to mid-June. The tuna turns purple upon maturing. Occasionally the cladophylls are purple. The Latin name, *Opuntia*, is derived from Opus, an ancient city in Locris, Greece.

Opuntia leptocaulis
TASAJILLO, SOB CACTUS   Color page 10, No. 4
A tall (2 feet) thin-stemmed cactus with long spines. Often it grows among other plants. Its flowers are greenish-yellow and the obovoid tuna is red (leading to the name Christmas cactus). It is found primarily in mesquite-short grass pastures.

Opuntia imbricata
CHOLLA   Color page 10, No. 5
Largest of the cacti found here and can attain a height of 4-6 feet. The flowers are reddish-purple and the tuna is yellow. It blooms from the first of May until the last of June. This conspicuous cactus can be found in grassland pasture and mesquite-short grass prairie.

Echinocactus texensis
DEVIL'S HEAD, HORSE CRIPPLER Color page 10, No. 6
A low-growing dome-shaped cactus with large, thick and flattened spines that curve downward. The plants usually are solitary and can be 8-10 inches in diameter and up to 4 inches tall. The flowers vary from nearly white

through pink to red. The flowers begin to open in late morning and reach a peak near 2:00 pm and close in late afternoon. Flowering begins in late April and extends through mid-May. The fruit is pulpy and turns a bright red; the seeds are large and black. Found in open grassland and is especially abundant in pastures. The generic name is derived from Gk., *echinos*, hedgehog, and *cactus*.

Echinocereus reichenbachii
HEDGEHOG CACTUS   Color page 11, No. 1
A short (up to 8 inches) cactus with large pink, red to purple flowers appearing briefly from late April to mid-May. The flowers open in late morning and close in late afternoon. It is found primarily in rocky, exposed caliche areas or in mesquite short grass pastures. Most abundant along the slopes of entrenchment canyons. The generic name is derived from Gk., *echinos*, hedgehog, and Lat., *cera*, wax.

132 LYTHRACEAE: LOOSESTRIFE FAMILY

Leaves simple, entire, variously attached. Fruit a capsule, one to several seeded. Contains the common ornamental crepe myrtle.

Lythrum dacotanum
LOOSESTRIFE   Color page 11, No. 2
Found along the edges of playas and other wet areas such as ditches. The small red-purple flowers appear in early June, peak in July and trail out into September, much dependent upon precipitation schedules. This is a tall (2 to 3 feet) but spindly plant that usually is found in a near-horizontal position. The generic name is derived from Gk., *lythron*, black-blood, and probably refers to the deep-purple flowers or perhaps because the plant turns black on drying.

Ammannia auriculata   Color page 11, No. 3
A small-flowered plant inhabiting playas. The tiny purplish flowers are in leaf axils of the primary stem of the willowy plant. Not overly familiar to us, it seems to flower in late summer and fall.

134 ONAGRACEAE: EVENING PRIMROSE FAMILY

Leaves variable. Flowers 4- or 5-merous. In ours, the stigma deeply 4-lobed. Ovary inferior, fruit an elongated pod. Seeds numerous. Primrose means firstling of spring.

Oenothera
EVENING PRIMROSES   Color page 11, No. 4
                               Color page 11, No. 5

There are numerous members of this genus on the Llano Estacado. Some emerge from the ground from a vast tap root with few leaves and a single, large yellow flower; others are bushes that literally are covered with small yellow flowers. One plant, probably (*O. jamesii*), observed in Los Lingos canyon stood over five feet tall.

The plants are most frequently associated with soils laden with caliche. Road cuts and entrenchment canyon slopes are prime habitat. *Calylophus serrulata* (a close relative to the oenotheras) blooms during the day. This bush (6 to 18 inches tall) is common and in full bloom is most beautiful. *Oenothera missouriensis* or flutter mill is found also in caliche soils and opens in late afternoon; however, there is a large-flowered *Oenothera*, which is similar to the flutter mill, which opens in mid-morning. *Oenothera canescens* is a small-flowered plant that occurs primarily in moist places, especially around playas. Its red splotched petals are distinctive. The evening primroses blossom most of the spring, summer and fall. The generic name is derived from Gk., *oino*, wine, and *thera*, to aid or help.

Gaura villosa
GAURA   Color page 11, No. 6

A low bush-like plant 1 to 1 ½ feet in height. Found in towns along sidewalks, alleys and in vacant lots. Present in open pastures, prairies, sandhills, playas and along roadsides. The intricate pinkish flower is difficult to describe and you must look at it closely to appreciate its beauty. It flowers from mid-April through September but peaks in May. The generic name is derived from Gk., *gauros*, majestic.

Gaura parviflora
LIZARD TAIL GAURA　　　　　Color page 12, No. 1
A tall (up to 5 feet) plant that is common along roadsides. The flowering stem tapers to a long, drooping tip that resembles a limp, lizard tail. The flowers are small and reddish. This plant is conspicuous along roadsides and blooms most of the summer.

151 ASCLEPIADACEAE: MILKWEED FAMILY

Leaves opposite or whorled. Sap milky. Corolla with inner corona arising from base of central stamen column. Pollen grains united into paired pollinia carried from flower to flower on the tarsi of insects. Seeds silky appendaged and airborne. Asclepias was the god of medicine and good health in Greek mythology.

Asclepias latifolia
GREEN MILKWEED　　　　　Color page 12, No. 2
A thick-leaved, stout plant about 12 inches high. Its yellow-green flowers are clustered in upper nodes and appear in early May into September, although flowering peaks in June to July. Found along roadsides, in pastures and sandy areas. *Asclepias oenotheroides* occurs in the area although its distribution is not well known. It is less conspicuous than *A. latifolia*.

Asclepias engelmannia
ENGELMANN'S MILKWEED　　　　　Color page 12, No. 3
This tall (up to 4 feet) slender plant with opposite, linear-lanceolate leaves has orange flowers. It blooms from mid-June into August. The plants are solitary and are observed sporadically, usually along roadsides at the edge of fields.

152 CONVOLVULACEAE: MORNING GLORY FAMILY

Twining vines. Flowers 5-merous, the corolla trumpet-shaped and showy, except in the parasitic dodder. Fruit a dry capsule of 1 to many seeds. The beautiful flowers open in mid-morning from mid-April into September. Quite spectacular in May-August. This family includes the sweet potato. The name is derived from Lat., *convolvo*, I entwine, or *convolvere*, to roll up.

Convolvulus arvensis
BINDVINE   Color page 12, No. 4
Morning glories are often found in what appears to be large mats along the roadsides. The vines are long (up to 3 feet). The corolla is funnelform, white or occasionally pink with broad vertical bands on the outside.

Convolvulus equitans
BINDVINE
Similar to *C. arvensis* but the corolla has a narrower tubular base, is 5-angled with the angles often projecting as points. The flowers are white to pink, frequently with a red throat. The bindvines are found primarily along roadsides and in open fields. They grow in compact soils and often reach high densities detrimental to cultivated plants.

Cuscuta
DODDER, ANGEL'S HAIR,
DEVIL'S GUT   Color page 12, No. 5
Has the appearance of a wad of spaghetti growing as a mat on top of a variety of plants. It is parasitic and lacks significant amounts of chlorophyll. The stems are yellowish-brown. The inconspicuous white flowers can be profuse and are observed during mid-summer. Dodder refers to weak, trembling — to doddering. The generic name is derived from Arabic, *kushuth*.

153 POLEMONIACEAE: PHLOX FAMILY

Gilia rigidula
BLUE GILIA   Color page 12, No. 6
A perennial herb with sepals partially united. Corolla is funnelform with 5 prominent and showy lobes. This is one of the most beautiful flowers in the area. The petals are violet-blue with a deep yellow throat. The dime-sized flower opens in late morning and closes in late afternoon. The plants are found primarily along road cuts in the entrenchment canyons but are seen also on the upper surfaces. They are 2-10 inches in height with a single stem or multiples from a common base. It blooms from

mid-April through May with a second spurt in August and September if water is adequate. The generic name is in honor of Gilio, a Spanish botanist. The family name is derived from Gk., *polemos*, war. Apparently Pliny reported that a dispute over the discovery of the initial genus of the family, *Polemonium*, led to war.

154 HYDROPHYLLACEAE: WATERLEAF FAMILY

Leaves variable. Inflorescence 5-merous, the petals united. Flowers in scorpioid cymes (*Phacelia*) or simple (*Nama*). Fruit a capsule with few to many seeds.

Phacelia congesta
BLUE CURLS　　　　　　　　　Color page 13, No. 1
Found along the steep slopes just below the caprock rim in soils heavy with caliche. The blue flowers are seen from mid-April through June. The annual or biannual herb can be 4 to 12 inches in height. The generic name is derived from Gk., *phakelos*, a bundle, referring to the position of the flowers.

Nama hispidum
TINY 'TUNIA　　　　　　　　　Color page 13, No. 2
This small plant (6 inches in height) has small blue-purple flowers that appear in late April through August. The plants thrive in compact soils in open fields heavy with caliche. It is common but inconspicuous and the small flowers that resemble petunias superficially are often overlooked. The generic name is derived from Gk., *nama*, stream.

155 BORAGINACEAE: BORAGE FAMILY

Leaves simple, alternate. Corolla 5-lobed. Fruit breaking into four 1-seeded lobes. Flowers regular, cymose. Borage is probably derived from medieval Lat., *borrago*, from Arabic, *abu-rashsh*, father of sprinkling, source of sweat. Medicinally, some European species were used as demulcents or diaphoretics.

Lithospermum incisum
APACHE TEA, PUCCOON Color page 13, No. 3
This delicate lacy-edged yellow flower is among the earliest to appear in the new year. Flowering begins in late February and ends in early May. The plant is about 7 inches in height and flourishes in hard packed caliche soils. It is found along roadsides in the sand hills. Tea brewed from this plant was thought to serve as a birth control measure. Interestingly, several hormone-like molecules have been identified from its juices: one inhibits the action of gonadotrophic hormones and one lowers blood thyroid hormones. Both actions make pregnancy more difficult to achieve. The generic name refers to rock seed.

Heliotropium
HELIOTROPE Color page 13, No. 4
Members of this genus are rare on the Llano Estacado. The 6-8 inch-high plants are spotty in occurrence but are found near the rim. We have seen them in bloom in April and May. The flower is distinctive; the five white petals are united to form a nearly flat face with a small constricted tube. A tall heliotrope (2 feet) is found in the sandy areas to the east of the Llano Estacado. The generic name is derived from Gk., *helios*, the sun, and *tropos*, to turn, and depicts the plant's tracking of the sun's path.

157 VERBENACEAE: VERVAIN FAMILY

Stems 4-angled, erect or prostrate. Leaves opposite, variously lobed or dissected. Inflorescence in axillary or terminal spikes or racemes, loosely or densely flowered, becoming greatly elongated in fruit. Corolla regular to slightly irregular. The verbenas are a large and diverse group. Traditionally, the prostrate forms are called verbenas and the erect ones vervains; some are trees. The names are from Lat., *verbena*, sacred boughs.

Verbena bipinnatifida
PRAIRIE VERBENA Color page 13, no. 5
A low-growing plant with many stems from a deep root. The 12-inch long stems turn upward and the lavender to purplish perfect flowers occur in bracted pedunculate

spikes. The leaves are opposite and bipinnatified. Flowers from late March through August. Spotty in occurrence and rarely in dense stands.

Verbena brachiata
VERBENA   Color page 13, no. 6
A low-growing verbena with many stems radiating from the central stock. Initially prostrate, the 6-8 inch stems turn upward and are capped with an expanded peduncle often ringed with tiny, bluish flowers. Common in towns, it is found in lots, alleys and along roadsides in compact soil. It flowers throughout the summer and into the fall but its beauty diminishes late in the year. *Verbena pumila* might also be in the area.

Verbena neomexicana
VERVAIN   (Not pictured)
This vervain grows to a height of 10-12 inches. Several stems emerge from a single stock, much like the arms of a saguaro cactus. The small purple flowers often go unnoticed. It grows in flat areas with some moisture and usually is obscured by grass. It flowers from May to midsummer. *Verbena plicata* might occur in the area. We have found that it is best to simply enjoy the vervains and not to worry too much about exactly what they should be called!

Phyla incisa
FROG FRUIT   Color page 14, no. 1
Found rarely along roadsides, around playas and parks. This trailing perennial herb usually has purplish stems and bright green, opposite leaves. The flowers are white with a red dot, grow in dense, cylindric spikes that are brown and can become an inch long. Flowering extends from early May through September. A similar plant, *Lippia*, is sold as a ground cover and is noted for its toughness. Its presence in towns reflects that it is an escapee. The generic name is derived from Gk., *phylon*, a race or tribe; a genetically related group.

158 LAMIACEAE (LABIATAE): MINT FAMILY

Stems square, leaves opposite. Most species aromatic. Calyx more or less irregular. Corolla two-lipped, highly irregular. Stamens 2 or 4. Many species can be used as herbs or condiments for flavoring; witness iced tea with mint and the mint julep. The older family name is derived from Lat., *labium*, lips.

Lamium amplexicaule
HENBIT Color page 14, no. 2
An early spring bloomer beginning in the first of March and continuing through June. This is a common pest plant in lawns and is rarely found outside of municipal areas. The flowers are rose-purple with dark spots inside the tubes. The generic name is derived from Gk., *lamios*, throat, and refers to the shape of the corolla.

Monarda citriodora
LEMON BEEBALM, HORSEMINT Color page 14, no. 3
Patches of this striking plant are seen scattered throughout the area. The pink flowers with a dash of purple are seen from late April through May and early June. If the precipitation schedules are in accord, occasionally it will bloom into July. Usually occurring in flat places, it is found in a variety of soils but more often than not it will be found in deep grass or at the edge of taller vegetation. The individual plants stand 8-10 inches high. The genus is named for M. Monardez, a physician who lived in Seville.

Teucrium laciniatum
GERMANDER Color page 14, no. 4
This mint with its white flowers grows to about 5-6 inches in height. It flowers from late April to mid-June, in flat places, primarily along roadsides, pastures and in some open prairie. Named for a king of Troy, Teucer.

159 SOLANACEAE: POTATO (NIGHTSHADE) FAMILY

Leaves alternate, simple to odd-pinnate. Calyx 5-toothed. Corolla tubular, 5-lobed. Fruit a berry or capsule. Contains such plants as tomatoes, potatoes and peppers as well as the dreadful tobacco plant. Some species are poisonous to livestock others hazardous to one's health!

> Tobacco is a filthy weed,
> From the devil doth proceed.
> Robs your money,
> Burns your clothes,
> And makes a chimney of your nose.
>   Anon.

Physalis viscosa
GROUND CHERRY   Color page 14, no. 5
This leafy green plant is found throughout the area. It excels in hard packed soils with or without caliche, especially if the soils have been disturbed. It is common in municipal areas, along sidewalks, vacant lots and along roadsides. The small yellow flowers with brown centers face downward and are often missed. It blooms from mid-April to mid-September and the flowers open in the morning. The plant can reach a height of 10-15 inches. The generic name is derived from Greek and means bladder or inflated pod.

Quincula lobata
PRAIRIE LANTERN   Color page 14, no. 6
One of the most beautiful flowers, its quartersized corolla is deep purple. It blooms from early April through September and opens in late morning. The plant is low-growing, often sprawling along roadsides and around playas. It is common in municipal areas. Closely related to *Physalis* but differs in flower color and by the fact that the flowers of *Physalis* face down and are hidden from view. The flowers of *Quincula* face upward and are conspicuous. Knowing the weakness of our reasoning in assigning species to genera, we will throw in author's choice, an euphemism for personal prejudice!

Solanum eleagnifolium
TROMPILLO, SILVER-LEAF NIGHTSHADE
                                      Color page 15, no. 1
The plant is considered a nuisance plant in municipal areas as well as to farmers. It is a perennial, springing from a deep tuber and is conspicuous along the shoulders of roads. The washed-out purplish flowers with prominent, yellow anthers are seen from early May into mid-September. The plants are about 12-15 inches in height

and the leaves are a light green, sometimes grayish. The fruit, a berry usually smaller than a marble, varies from light green with dark green stripes, to yellow and finally to brown. It is poisonous to livestock. The leaves of *S. dimidiatum* (western horsenettle) are irregular lobed, broadly lanceolate to ovate while those of *S. eleagnifolium* are lanceolate and generally unlobed. *Solanum eleagnifolium* is found throughout and *S. dimidiatum* might be more prevalent to the south. The generic name is derived from Lat., *solanem*, comforting, and refers to the sedating qualities of some of the plants. *Solanum* means nightshade in Lat.

Solanum triquetrum
WHITE NIGHTSHADE   Color page 15, no. 2
This tall but spindly annual can attain heights of 2-3 feet. It has tiny white flowers with conspicuously yellow pollen-producing structures, and blooms from June to August. It is found in municipal areas, in unkempt flower beds and under bleachers! It seems to require some shade. The small berry is bright red.

Solanum rostratum
BUFFALO BUR   Color page 15, no. 3
This distinctive plant with its "watermelon" leaves is covered with sharp spines — even its seed capsule looks like a cocklebur! The plants are about 12-15 inches in height and grow in nearly identical situations with trompillo. The yellow flower has large, curved stamens. Flowering begins in early June, noticeably later than trompillo, and continues through September. A perennial, it is considered by most to be a pest species.

Chamaesaracha coronopus
FALSE NIGHTSHADE   Color page 15, no. 4
This much branched plant with highly dissected leaves thrives in sidewalk cracks! It is found in municipal areas, and along roadsides in disturbed but hard-packed soils. The washed yellow flower opens in mid-morning and closes in late afternoon. Flowering extends from mid-June through mid-September. This is a nice, quiet plant. The generic name is derived from Gk., *chamai*, humble, on the ground, meaning low, which it certainly is.

Datura wrightii
JIMSON WEED Color page 15, no. 5

A most impressive plant, introduced. It occurs in municipal areas in alleys and as an ornamental. It can be seen around barns and stock pens. This large, tall plant (3 feet) has striking trumpet-like, creamy white flowers that can be 5-6 inches in length. They open in early evening, are quite sweet smelling and are pollinated primarily by hawk moths. The walnut-sized spiny seed capsules have a strong odor and are affectionately called "porcupine eggs." The plants are poisonous and the foliage has an unpleasant odor — in fact, it stinks!

At Jamestown, soldiers made a salad to which they added the tender, young leaves of *Datura*. Shortly, several of the soldiers began to "behave badly" throwing at and chasing objects that did not exist; talking in garbled sentences; stripping off their clothes and frolicking through hill and dale; kissing and stroking other soldiers; and, in general, acting as though they were at a college fraternity party. The plant was called Jamestown weed, which has been shortened through time to jimson weed. The generic name is derived from Sanskrit, *dhattura*, thorn apple.

160 SCROPHULARIACEAE: FIGWORT FAMILY

Corolla mostly irregular, generally bilabiate. Fruit a 2-celled capsule with many seeds. Flowers in racemes or panicles. Scrofula is a morbid condition that manifests itself in an enlargement and degeneration of the lymph glands, especially those in the neck. The swollen glands were fancifully likened unto little pigs, i.e., *scrofa*, Lat. for breeding sow. Certain of these plants were thought to cure or arrest the condition.

Penstemon
BEARD-TONGUE, PENTSTEMON Color page 15, no. 6

These 12-15 inch-tall plants are found primarily along road cuts associated with caliche. Occasionally they will be found in mesquite-short grass pastures. The white to lavender flowers are observed from mid-April through May. The name beard-tongue refers to the sterile, hairy stamen found in some species. The generic and common names refer to five stamens. Note that the common and generic names are spelled differently.

Castilleja sessiliflora
INDIAN PAINTBRUSH　　　　　Color page 16, no. 1
This plant is nearly restricted to the canyon slopes, steeped in caliche. The yellowish-green to pink flowers are not nearly as striking as its more southern cousin. More often than not it is overlooked even though the plant is about 6 inches tall. Flowering extends from the first of April through May. You will have to search this one out! The genus is named for a Spanish botanist of the same name.

162 MARTYNIACEAE: UNICORN PLANT FAMILY

Proboscidea louisianica
DEVIL'S CLAW　　　　　Color page 16, no. 2
Stems prostrate, to 3 feet long. Leaves and stems viscid-pubescent and ill-smelling. Flowers showy, somewhat bilabiate. Fruit a large capsule terminating in a long, incurving hook, separating as it ripens. The young fruits are edible and can be pickled like okra. Found in open fields and pastures, generally in loose soils. This large-leaved plant flowers from the last of June into September. The 5 petals are united into a 2-lipped corolla about 1 inch in length. The flowers are nearly white to lavender with yellow or purplish spots. Note the dark stripes emanating from the throat; they appear to be stamens at first glance. The fetid odor of this plant and its flowers is enough to assure you that one close look might be adequate. However, in other parts of the world the plant is grown for its green pods, which are cooked and preseved much like okra. The opened and dried pods are used decoratively. The generic name is derived from Gk., *proboscis*, a snout, referring to the shape of the fruit. The family is named for John Martyn, an English botanist.

168 RUBIACEAE: MADDER FAMILY

A large family that includes coffee and guinine. The leaves are opposite, simple, with sheathing stipules. Corolla tubular, with 4 spreading, showy lobes. The family name is derived from Lat., *ruber*, red, and refers to the dye extracted from the roots of some of the plants. Alizarin is the dye but it is synthesized today.

Hedyotis nigricans
STAR VIOLETS Color page 16, no. 3
This star violet is a low bush-like plant (about 1 foot high) with few to a large number of stems arising from a woody stock. It is found primarily on hard-packed caliche soils and in pastures. The small cream-colored to pinkish flowers cover the tops of the plant. It flowers from late May into June but the show is over before July. We have found it on the Llano Estacado associated with entrenchment canyons. The generic name is derived from Gk., *hedy*, sweet, pleasant, and *osis*, condition.

Hedyotis acerosa
STAR VIOLETS Color page 16, no. 4
This star violet grows on the bare caliche slopes. It is wide spread but is often overlooked because of its size; usually less than five inches in height. The flowers are larger, though more sparse, than those of *Hedyotis nigricans* and usually show more tendency toward pink. Viewed closely, the four-lobed flower is seen to have a long corolla tube and the flower is quite hairy. The leaves are coarse and narrow and the plant usually emerges from a single tap root. It flowers throughout the summer and early fall.

171 CUCURBITACEAE: GOURD FAMILY

Annual or perennial from a large to very large, deep tuber. Stems prostrate and long. Leaves entire or variously divided. Flowers dioecious, corolla yellow. Fruit a fleshy, hard-rinded, many-seeded berry. Yes, berry!

Cucurbita foetidissima
BUFFALO GOURD, STINK GOURD Color page 16, no. 5
A striking plant with long (6-8 feet) vines radiating from a central woody stem. The greenish-gray leaves are large,

triangularovate and simple. The plant has a stifling, fetid odor which lingers. The flowers are bright yellow, opening in late morning. Flowering begins in early May and extends into July. The large round gourd is green with dark green stripes; it fades to a uniform brown as it dries and the outer shell becomes papery. Conspicuous along roadsides, railroads, and in sandy areas and pastures. In various places around the world this plant is cultivated for the high quality oil in its seeds and the high concentration of carbohydrate in its tuber, which can be 8 inches in diameter and 2 ½ feet in length. One day when you are idle, dig one up! The generic name is derived from Lat., *cucurbita*, gourd.

Citrullus lanatus
PIEMELON, WATERMELON   Color page 16, no. 6
Of course these melons are introduced to the New World but they are found growing wild in the sand hills and in sandy areas, especially around what used to be the Bloated Goat Liquor Store near Bledsoe. The long vines emerge from a single woody stem. The flowers are a washed out yellow. Flowering begins in June. You can tell piemelons from watermelons by driving over the fruit in a truck: if the melon remains intact, it is a piemelon! The generic name is probably derived from Lat., *citrus*, and probably refers to the green-yellowish color of the fruit.

Ibervillea lindheimeri
BALSAM GOURD, GLOBE BERRY Color page 17, no. 1
The vine springs from a not so deep tuber that resembles a half-flattened ball, about 6-10 inches in diameter. The plant is rare on the Llano Estacado but can be found near the rim usually in mesquite-short grass pasture. It intertwines into bushes such as juniper, and unless it is blooming or has set fruit, it is difficult to separate from another vine, *Cissus*, which is unrelated. The greenish flower with stripes is inconspicuous but present in late May and June. The distinctive characteristic is the roundish, bright red gourd that is slightly less than an inch in diameter and resembles a Christmas bell. The generic name is derived from Lat., *Iberia*, an ancient

name for the Spanish peninsula, and *ville*, denoting a city, town or place.

172 CAMPANULACEAE: BLUEBELL FAMILY

Lobelia cardinalis
CARDINAL FLOWER　　　　　　　Color page 17, no. 2
Leaves alternate, entire, spirally arranged. Ovary inferior, sepals and petals 5-merous. Corolla irregular. Fruit a many-sided capsule. The cardinal flower is rare in the area. We have seen it only on the cliff faces around pools just off the Llano Estacado. The bright red flowers were observed in September and early October. This plant should be looked for in the deep, moist canyons that dissect the eastern front of the Llano Estacado. The cardinal flower is the only member of the family that is not blue. The name is derived from the bright red array worn by Cardinals in the Catholic church. The genus is named for M. Lobel, a botanist and a physician to James I. The family name is derived from Lat., *campana*, bell.

174 ASTERACEAE (COMPOSITAE): SUNFLOWER FAMILY

Ah, yes! The composites! This dazzling and mind-boggling family of plants will provide you at once with the most beautiful display of flowers and the most supreme of challenges. This is a large family having in common the characteristic that the flowers are borne in dense heads on a single receptacle subtended by one or more whorls of bracts, the whole appearing as a single flower. *Aster* is derived from Gk., *astrom*, a star.

Vernonia
IRONWEED　　　　　　　　　Color page 17, no. 3
The purple, profuse heads of this 2 ½ feet-tall plant can be seen in colonies in early July into August. It is found primarily in moist areas in ditches along roadsides and around some playas. A complex group, the ironweeds obtain their name from their resistance to pulling and the toughness of their stems. A nice plant, often overlooked! The genus is named for W. Vernon, who collected and

sent plants from Maryland to James Petiver, an English naturalist.

Liatris punctata
GAY FEATHER   Color page 17, no. 4
This plant is found primarily on roadside cuts, canyon slopes and short grass prairies heavy with caliche. Its 6-8 vertically sweeping stems radiate from a common base and can attain a height of 2½ feet although usually shorter. A perennial with a thick, woody root. The willowy, wine-purple flower heads can be seen from late August through September. The origin of the generic name is unknown.

Kuhnia
FALSE BONESET   Color page 17, no. 5
Little is known about this plant on the Llano Estacado. It emerges from one to a few stems that spread laterally then sweep upward. The plant can be 1 ½ feet high but is usually shorter. It is found on hard-packed soil, usually associated with caliche on the slopes of entrenchment canyons. The disc florets are cream-colored and ray flowers are absent. One of our latest bloomers, it begins in late September and goes through mid-October. When seed has set, the plant is literally covered with white "puff balls" that remain deep into the winter. The genus is named for Adam Kuhn of Philadelphia. Boneset refers to members of the genus *Eupatorium*, which have reputed diaphoretic qualities. In folkmedicine, the perfoliate leaves (the stems appear to penetrate the leaves because the bases of the leaves are fused around the stem) were thought to signify some quality of bone healing. Thus, they were placed in the wrap over the bone.

Heterotheca
CAMPHOR WEED   Color page 17, no. 6
The yellow flowers of this plant form a head varying in size from a dime to a nickel. A low, bushy species is rare but an erect, spindly species is common. Generally around 2 feet in height, the sandy soils in the south nurture individuals over 4 feet tall. Primarily found in pastures, along roads at the edge of fields or in fallow land. Its distinctive odor of menthol and the late July through

October flowering period cause one to associate it with the fall of the year. The generic name is derived from Gk., *hetero*, other, different in kind, and *theke*, a case to put something in, referring to the different forms of the pollen sac of an anther.

Xanthisma texanum
SLEEPY DAISY   Color page 18, no. 1
This 1 to 1 ½ foot plant is so named because its flowers open in early afternoon after what appears to be a laborious effort to do so! The inflorescences are usually larger than a quarter when fully opened and the 22 or so long, yellow ray flower petals are bright and busy. The disc is also yellow. Found along roadsides, open range, fields and vacant lots. Flowers from early May through August. The generic name is derived from Gk., *xanthisma*, that which is dyed yellow.

Grindelia squarrosa
CURLY-CUP GUM WEED   Color page 18, no. 2
This is a bushy plant with bright green, relatively small leaves. It is usually found in sandy, moist areas along roadsides, in pastures and along fence rows. It is found in disturbed caliche, the plants are usually quite small. Normally it attains a height of 2 ½ feet. The yellow flowers are about ¾ of an inch across and the bracts and stems exude a sticky resin. It blooms from early August throughout the fall. The genus is named for H. Grindel, a German botanist.

Prionopsis ciliata
SAW-LEAF DAISY   Color page 18, no. 3
This is a tall, stout annual that can reach 4-5 feet. The leaves have coarse teeth. Considered a pest plant, it is common in dense stands in vacant lots, fields, pastures and along roadsides. Its yellow-flowered heads are terminal on a few short branches at the tip of the plant. It flowers from mid-June, doesn't peak until the end of July-August, and trails into September. The generic name is derived from Gk., *prion*, a saw, and *opsis*, appearance, referring to the leaves.

Gutierrezia sarothrae
BROOMWEED, SNAKE WEED   Color page 18, no. 4
This is a perennial plant that is quite dense in overgrazed pastures or range land. The green foliage is dome shaped, is usually 12-20 inches high and can become nearly covered with the small yellow flower heads in September. Due to its high resin content the plant is easily sparked to flame when green and is a veritable torch when dried. Some species are poisonous to livestock but judging from its dense stands, must also be unpalatable.

Haplopappus spinulosus
HAPLOPAPPUS   Color page 18, no. 5
This low-growing plant appears to be gray-green. Its ray (22-24) and disc flowers are yellow. Flowering from mid-April into September, the plants thrive in hard-packed soils with or without caliche and are found in fields, vacant lots, pastures and along roadsides. This plant is known for its small number of chromosome pairs (4) and is used as a scientific model to investigate chromosome "behavior." The generic name is derived from Gk., *haplo*, meaning single or simple, and *pappus*, meaning an old man, referring to the single row of bristles on the achenes. Originally described as *Aplopappus*, but was later corrected to *Haplopappus*. Note also the origin of the noun, *papa!*

Isocoma wrightii
JIMMY WEED, ROCEA   Color page 18, no. 6
A conspicuous plant, it is rare on the Llano Estacado. It has been observed on the rim near Post and is probably common on the eastern front. It is usually seen along roadsides and cuts in the deep red Triassic soils or in sandstone. It attains a height of about 12 inches and has many stems from a common base. The bright yellow flower heads appear in early July into August. Highly poisonous to livestock. The generic name is derived from Gk., *isos*, equal, and *kome*, hair. The inflorescences appear to be fuzzy.

Machaeranthera tanacetifolia
TANSEY ASTER, TAHOKA DAISY Color page 19, no. 1
The rays of this most beautiful annual are purple and the

disc is bright yellow. Usually about 10 inches high, it can attain a height of nearly 2 feet. There are 24-30 ray petals and the inflorescences are 1 to 1 ¼ inches across. Occasionally white! The plant thrives in flat areas with compact soils (with or without caliche) in vacant lots, along roadsides, pastures, short grass slopes and fields. Blooms from early April (peaks in May) into September. Sold through commercial outlets as Tahoka daisies. The generic name is derived from Gk., *machaira*, dagger, and *anthos*, flower, referring to the shape of the ray flowers.

Aster subulatus
ANNUAL ASTER   Color page 19, no. 2
First, let us say that there are several white asters with yellow discs; some are low-growing. The annual aster has small, white, numerous ray flowers and is a conspicuous floral component of playas and other moist areas. The disc florets are yellow and the ray flower petals take on a pink-bluish tinge with age. Plants attain heights of 2-3 feet and literally clog the playas in good growth years. It blooms from mid May into September, depending on precipitation. The generic name is derived from Gk., *astrom*, a star.

Melampodium cinereum
MOUNTAIN DAISY, ROCK DAISY Color page 19, no. 3
A conspicuous perennial that thrives in flat areas or slopes steeped with caliche. It is low to the ground and has narrow leaves. The inflorescence is nickel-sized and the 8 ray petals are white. The disc is yellow. Blooms from early April, peaks in late April to May and continues into September. A common, beautiful plant worthy of anyone's attention. The meaning of the generic name is often misinterpreted (black foot). It honors Melampus who, in Gk. mythology, was a seer having the gift to talk to animals and plants.

Berlandiera lyrata
GREEN EYES   Color page 19, no. 4
This plant is usually less than 12 inches in height and is sporadically common along roadsides, in pastures, vacant lots, and fields. It tolerates compact soils with much caliche and arises from a leafy mat. The 8 or so ray

flowers are yellow-orangish, with maroon veins beneath. Disc flowers have maroon corollas and the bracts are large, shield-like and overlapping. The ray flower petals roll up during the day and open in early morning if water is available. Flowers from mid-April through September. The ray flowers and most of the disc floret corollas will drop off, leaving the gray-green receptacle with the large shield-like bracts. The scattered disc floret corollas appear as eye lashes. Named after Jean Louis Berlandier, one of Texas' first naturalists.

Engelmannia pinnatifida
CUT-LEAF DAISY Color page 19, no. 5
This is a dense perennial with one or two 12-inch stems from the rosette-like base. The leaves are deeply dissected and the 8 or so unnotched ray flowers are yellow, as is the disc. The lower ⅓ of the disc anthers are brown. The ray flower corollas curl backward for part of the day or during dry spells. The receptacle is hairy, with 20 or so long narrow flaps. This is a common plant often found in conjunction with *Berlandiera* in compact soils. It is found along roadsides, vacant lots and fields. Flowers open in early morning from mid-April through September.
Henry Shaw was an English emigrant who, feeling great concern that he do something philanthropic for the peoples of St. Louis, Missouri, began development of a ten-acre garden like the great estates of Europe. Dr. George Engelmann, an early, great American physician-botanist persuaded Shaw to include a herbarium and library to supplement the garden. The Missouri Botanical Garden was opened to the public in 1859 and remains one of the finest display/research herbaria in the world.

Ambrosia
RAGWEED Color page 19, no. 6
Several species of this medically important genus are found in the area and many of the problems associated with hay fever and allergies are attributed to their pollen. *Ambrosia grayi* is commonly found in moist areas, notably around playas. It attains a height of 12-18 inches and the highly dissected leaves are silver-gray. Ray flowers are absent and the heads are unisexual but both the stam-

inate and pistillate flowers are found on the same plant. Staminate flowers occur near the top, pistillate flowers below. The flowers hang down like bells. Flowering occurs from mid-summer through the fall. The generic name is derived from Gk., *ambrotos*, immortal. Ambrosia, along with nectar, formed the food and drink of the ancient gods. To partake was to become immortal.

Zinnia grandiflora
ZINNIA   Color page 20, no. 1
A low-growing (6-8 inches) plant with a single main stem. The plants grow in clumps, usually in flat areas; along roadsides, in pastures, vacant lots and open fields. The inflorescence is distinctive with 3 dull yellow ray flowers. The receptacle is narrow. Leaves narrow, often curling. The ray flowers are persistent, lose their color and take on a parchment-like character. Flowers late in April with a peak in May but trails into September. Named for J.G. Zinn, a German botanist.

Ratibida columnaris
MEXICAN HAT   Color page 20, no. 2
A distinctive plant with its 1 to 1 ½ inch brown-green cone surrounded by 5-8 ray flower petals. The petals can be uniform yellow to a deep maroon edged in yellow. It grows in flat places, along roadsides, in fields and lots. Generally less than 12 inches high. Flowering begins in late April, peaks in mid-late May but continues into September. However, the individual flowers will be uspectacular late in the season.

Helianthus annuus
PRAIRIE SUNFLOWER   Color page 20, no. 3
The typical annual sunflower that can stand up to 6 feet. It is found along roadsides, in fields, lots and pastures. A cultivated form with a head 10 inches across is a common escapee. *Helianthus maximiliani* has been found only in the deep entrenchment canyons around pools. It is identified by its narrow leaves. The sunflowers bloom from early May through the fall. The generic name is derived from Gk., *helios*, sun, and *anthos*, flower.

Helianthus ciliaris
BLUE WEED Color page 20, no. 4
This perennial plant is common in disturbed areas with or without caliche. Considered a pest plant, it thrives in vacant lots, along roadsides and in open fields where the soil is compact. It attains a height of 13 inches; its 11 or so ray flowers are yellow and the thick conspicuous disc is brown with yellow-tipped anthers. Flowers from mid-May, peaks in June and trails into September.

Verbesina encelioides
COWPEN DAISY Color page 20, no. 5
A conspicuous greenish-gray plant that attains a height of 4 feet but is usually 1-2 feet. It is found sporadically in disturbed areas along roadsides, in fields and in pastures. Generally, saw-leaf daisy is close by. The highly branched plant will have numerous flower heads, each with a yellow disc and about 13 (12-16) yellow rays that have three lobes at their tips. The disc is thick and becomes sprinkled with brown as the heads age. Flowering begins in July and continues into September. A plant that just suddenly seems to appear! A close relative, *Verbesina virginica*, is called the ice plant for its sap freezes in such a manner as to form broad sheets of ice that burst and grow from the plant. The juices extracted from this plant are used to treat a variety of internal ailments. The generic name is derived from verbena.

Coreopsis tinctoria
COREOPSIS Color page 20, no. 6
This delicate annual has around 8 bright yellow ray flowers that are 3 lobed. The inner portion of each ray is maroon and the combined effect is a maroon ring around a smallish, round, brown disc that is flecked with yellow. Found primarily in playas or in ditches. Flowers in May to June with a smattering of late summer activity. The generic name is derived from Gk., *koris*, a bug, and *opsis*, like, a reference to the seeds.

Thelesperma filifolium
THELESPERMA Color page 21, no. 1
This beautiful plant with its brown disc and yellow ray flowers is commonly in flower from mid-April to the end

of May but can be found throughout the summer. While it can be found throughout it reaches its highest densities in soils steeped in caliche. It attains a height of about 1 foot and its flower heads are about the size of a quarter. Generally it has 8 rays and 7-8 projections extending from the receptacle. Do not confuse with *Haplopappus spinulosus* which has 20-22 ray flowers and grows in dry areas.

Thelesperma megapotamicum
RAYLESS THELESPERMA   Color page 21, no. 2
One of the more bizarre appearing composites on the Llano Estacado. The 2-3 foot high plant has few branches and small inconspicuous leaves. At best, the plant could be described as spindly, yet it stands stiffly. The smooth involucre with 5 projections and absence of ray flowers and the yellow "fuzzy" disc make identification easy. A common plant, it is found in vacant lots, old fields, along roadsides and just about any place else. Possibly confused with the yellow old plainsman (*Hymenopappus*) that has a much stouter stem, larger leaves, far more heads and numerous individual bracts forming the involucre.

Gaillardia pulchella
FIREWHEEL   Color page 21, no. 3
This beautifully-flowered plant is about 6-10 inches high and is found along roadsides, vacant lots, pastures and other areas with sand. It does not attain the densities observed just off the Llano Estacado toward the SE. Each plant has one or two vertical stems with terminal heads. Most ray flowers are tipped with yellow; the disc corollas are yellowish below and purple or red above. Flowers beginning in mid-April, peaks in April or May and trails into July. The genus is named for Gaillard de Marentonneau, an 18th century French patron of botany.

Gaillardia pinnatifida
OLD RED EYE   Color page 21, no. 4
Attains heights of 12-15 inches and occurs in a variety of habitats, usually in flat areas along roadsides, open range, fields and lots. Occasionally observed on caliche road cuts. The ray petals are yellow, 3-pronged and have pur-

ple veins beneath. Blooms from mid-April, peaks in late April to May and trails into September.

Hymenoxys
BITTERWEED Color page 21, no. 5
This is a strong-scented plant of the caliche slopes. It is a perennial emerging from a stout taproot. The lanceolate basal leaves extend above ground about 3 to 5 inches; quite bushy. The heads with their yellow ray and disc flowers are terminal on slender 8- to 10-inch tall stems. The odor of the crushed foliage is similar to that of *Dyssodia*. Has a bitter taste. Flowers as early as January but March through April are peak flowering times. Trailing into May, the ray flowers can become parchment-like. The generic name is derived from Gk., *hymeno*, meaning membrane, and *oxys*, smooth.

Helenium badium
SNEEZEWEED, BITTERWEED Color page 21, no. 6
A taprooted annual or perennial with many terminal heads on numerous branches. It attains a height of 1 to 1 ½ feet and is dome-shaped or rounded in outline. The brownish-red disc, which might be sprinkled with yellow florets, is round and relatively large with regards to the head size. The 6-8 yellow rays are attached near the base and extend downward. Flowers from April into September but most spectacular in June to July. Found in moist areas around the shoulders of playas and in dry pastures, along roadsides and in lots. Another sneezeweed that is quite tall (3 feet) and unbranched has nearly round dull brown discs subtended by few small yellow rays. It grows in stands in moist areas, often in considerable shade. The generic name is from Gk., *helene*, basket, or *helix*, twisted or spiral.

Dyssodia acerosa
DOG WEED Color page 22, no. 1
A strong-scented perennial herb found primarily on caliche slopes, in rocky areas around canyons and in mesquite-short grass areas around the rim. A low-growing plant (5 inches) with yellow to light orange corollas. Blooms from early May until September. The generic

name is derived from Gk., *dysodia*, stench, referring to the fetid odor of the foliage.

Flaveria trinervia
CLUSTER FLAVERIA　　　　　　Color page 22, no. 2
An annual herb with opposite leaf blades, united at their bases around the stem. The leaves are toothed and the yellow (somewhat plain) flowers are clustered in the leaf axils. Sporadic in occurrence, it is found in hard-packed soil along roads, in open fields and pastures. Usually 1-2 feet high, it can exceed 3 feet. A pest plant in agricultural fields. Flowers from August through September. The generic name is derived from Lat., *flavus*, yellow.

Palafoxia rosea
SAND PALAFOX　　　　　　　　Color page 22, no. 3
This annual herb has a stout taproot and is erect, up to 18 inches tall. The pink flower heads are few and lack rays. Plants are found primarily in sandy areas with moisture. Canyon bottoms and ditches are acceptable habitats. Spotty in distribution and solitary. One species we have seen in the sand hills is short (6-8 inches), has single heads and has rays.

Psilostrophe villosa
PAPER FLOWER　　　　　　　　Color page 22, no. 4
A plant with numerous stems from a woody tap root. About 10-12 inches high. The bright yellow flower heads are terminal. Found in open fields, pastures, along roadsides, open range and vacant lots. Often in dense stands. Flowers from late April, peaks in May and trails into July. The generic name is derived from Gk., *psilo*, mere, bare, and *strophe*, twist or turn.

Hymenopappus tenuifolius
OLD PLAINSMAN　　　　　　　Color page 22, no. 5
This is a tall slender plant with numerous terminal inflorescences. It grows well in flat areas: in fields, pastures, vacant lots, and along roadsides throughout the area. Ray flowers are absent but the phyllaries can be tipped with white, resembling rays. The disc florets are creamy white. *Hymenopappus flavescens* is quite similar to *Hymenopappus tenuifolius* but the heads are yellow. The two species do not appear to coexist to any significant

degree. *Hymenopappus flavescens* has a strong, pungent odor. Both appear to be annuals but the rootstock of *Hymenopappus tenuifolius* might be persistent. The flowering schedules of these two species vary from year to year. Generally, the yellow-flowered species flowers earlier (late April-early May) but if growth and flowering are delayed, then the white-flowered species might flower first. Do not confuse the yellow-flowered species with rayless thelesperma. The generic name is derived from Gk., *hymen*, membrane, and *pappus*, referring to the tuft of appendages crowning the ovary or fruit.

Achillea millefolium
YARROW, MILFOIL   Color page 22, no. 6
This tall plant with fern-like leaves and 1-2 stems attains a height of nearly 2 ½ feet. A perennial, we have seen it growing only along the roadsides or within medians. Sporadic in occurrence, it might be mowed before it has enough time for the white ray flowers to appear. The disc can have a tinge of yellow. Flowers in July and August. Named for Achilles, a pupil of Chiron, who first used the plant as a medicine. A few leaves steeped in boiling water for a few minutes makes a pleasant tasting tea. We have tried it frequently!

Senecio douglassi
THREADLEAF GROUNDSEL   Color page 23, no. 1
A 2 to 2 ½-feet-high perennial found along roadsides in hard packed or sandy soil, often with caliche. The flower heads are relatively few, have long, narrow, yellow rays and are borne on drooping peduncles that exceed the leaves. The plants are usually isolated. Flowering starts in mid-April, peaks in late April to May. In late September another species, *S. riddellii*, blossoms furiously, but briefly. The generic name is derived from Lat., *senex*, an old man, probably referring to the naked (bald) receptacle.

Centaurea americana
STAR THISTLE   Color page 23, no. 2
This erect annual attains heights of 3 feet or so. Found in dense clumps generally associated with caliche slopes. The inflorescences lack ray flowers and the disc varies

from light pink around the outside through pink and white in the center. Inflorescences can be 4-5 inches across, usually smaller. Rarely all white! Flowers from the first of May into July. The generic name is derived from Lat., *centaureum*, from the centaur Chiron, who is reputed to have discovered its medicinal properties. Chiron was the most famous of the centaurs and achieved renoun for his knowledge of medicine.

Cirsium texanum
TEXAS THISTLE Color page 23, no. 3
A biennial or perennial herb that attains a height of 3 feet. It is found in patches, usually along roadsides or as isolates in pastures, around barns and feed yards. It lacks ray flowers and the involcre is cup-shaped. The disc corollas are a rosy-lavender. Flowering begins in late April, peaks in May and trails into July. A beautiful prickly plant often used in dry arrangements. Rarely all white! The generic name is derived from Gk., *kirsos*, a swollen vein. An apt description of the results of being pricked by the spines.

Perezia nana
DESERT HOLLY Color page 23, no. 4
Locally abundant in colonies associated with the juniper-agarita association (open range). Propagated by rhizomes, the short (6-8 inches) plant has firm leaves that have spikes around the edges. More often seen as brown and parched. The living leaves are greenish-gray. The small pink head occurs at the tip of the plant. Flowering is reported to be sporadic and blossoms might not be observed unless you will be willing to crawl amongst the "cascabels." We have found it in bloom each year since we first observed it, always in May. Found to the extreme southeast but was recently observed south of Tahoka in Lynn County.

Stephanomeria
SKELETON PLANT Color page 23, no. 5
Grows in flat places often in hard soils but usually associated with sand. Found in vacant lots rarely but is most frequently observed along roadsides between the ditch and the plowed field. An inhabitant of fence rows. The roundish plant has thin but dense stems, reduced leaves

and often is confused with a young tumbleweed. The small flower heads have 5-6 lavender ray petals. Often the flowers do not extend beyond the foliage but we have seen a few plants with the exterior brightly covered. Blooms throughout June and July, sometimes into September.

Lygodesmia texana
SKELETON PLANT Color page 23, no. 6
This perennial herb has a single nearly leafless stem emerging from the ground. Its large and delicate lavender inflorescence opens early in the morning and although it can remain open in the afternoon if moisture is adequate, the wind and hot weather take a heavy toll. Rare on the Llano Estacado, it is found along road cuts and in mesquite — agarita range land to the SE. It flowers from late May into July and is worth looking and stopping for! The generic name is derived from Gk., *lygodes*, flexible, referring to the willowy stems.

Lygodesmia pauciflora
SKELETON PLANT (Not pictured)
Apparently a rare plant restricted to the slopes of entrenchment canyons. The flower head is the size of a nickel and the beautiful lavender ray flowers are miniatures of *L. texana*. The near leafless plant has many secondary stems more like *Stephanomeria*. We have seen it blooming only in September but the presence of old seed heads indicates an earlier flowering period.

Tragopogon dubius
GOAT'S BEARD Color page 24, no. 1
The puff ball or seed head of this plant is a sphere up to 4 inches in diameter! A common weed that sends up a single stout stem, up to 2 feet from a deep taproot. The leaves are long and narrow and droop downward in a tuft. The yellow flowers with their green bract extensions open in the early morning and close usually by mid-day. It flowers from April through July. Dubiously edible! The generic name is derived from, *tragus*, goat, and *pogon*, beard.

Sonchus and Lactuca
SOW THISTLE AND WILD LETTUCE
Color page 24, no. 2
These two plants are similar in many respects, although

sow thistle is introduced from Europe and lettuce is native. Both have milky sap and grow in disturbed soils; along roadsides, in fields, alleys, lots and in sidewalk cracks! The inflorescence of each is similar, less than a dime across and milky yellow. Both flowers open in the morning. Probably *Sonchus* flowers earlier, beginning in March. *Lactuca* flowers in late June and July; however, there is considerable overlap. *Sonchus sends up a single, stout, rough-looking stem with few terminal branches. Sonchus is considerably more spiny than Lactuca.* In addition, the pappus bristles of *Sonchus* arise from the top of the seed whereas those of *Lactuca* are on the top of a slender stem. *Lactuca* is tall also but the stem is much more delicate and the terminal branches are long, spindly and numerous. Both can attain a height of 3-4 feet but such a *Sonchus* looks like a pine tree! Considered to be pest plants, some are edible (*Lactuca sativa* is lettuce) and are good fodder for ruminants and geese. *Lactuca* refers to Lat., *lactus*, milk. *Sonchus* is derived from Latin, sow thistle, from German, *sonchos*.

Taraxacum officinale
DANDELION Color page 24, no. 3
This introduced perennial weed has basal leaves and usually a single stem with a terminal bright yellow flower head that can be 8 inches above the ground. It is a common pest plant in disturbed habitats, primarily in municipal areas. You might not wish to consider your lawn a disturbed area but 'tis the dandelion that shall decide! It flowers most every month but has a peak period in the spring and early summer. A native of Europe, it is eaten in salads, or pickled, but most of all, and with great gusto, it is simply killed! Most lawn aficionados seem to reside under the umbrella of, "It can live here but it had better not bloom." Yet, it has a most exquisite flower. The original meaning of the generic name is obscure. Dandelion is derived from French, *dent de lion*, lion's tooth.

As with the dandelions, this book has sprouted, flowered and now it is time to set seed! PLEASANT JOURNEY!

Vocabulary of Botany

Botanists not only live in their own world but they also have their own language. Some of the terms do make sense, others you will just have to accept on trust. With a little practice and initial work you can quickly gather a working handle on the terms in order to more efficiently use the keys and descriptions. We have culled many terms but the following list is nearly essential.

Achene — A small dry, nut-like fruit.
Alternate — Placed singly on the stem (not opposite or whorled).
Anther — The part of a stamen that develops and contains pollen and is usually borne on a stalk.
Axil — The angle formed by a leaf and its stem.
Axillary — Arising from an axil.
Berry — A fleshy fruit containing few to many seeds.
Bipinnate — Twice, or doubly, pinnate.
Bract — A modified leaf subtending a flower or flowers (if several they form an involucre).
Calyx — Think of chalice. The external, usually green or leafy part of a flower consisting of sepals.
Capsular — Like a capsule.
Capsule — A dry fruit containing few to many seeds and opening by pores or slits.
Carpel — A simple pistil or one unit of a compound pistil.
Cladophyll — A flattened stem, assuming the shape and function of a leaf, a cladode.
Corolla — The petals of a flower.
Corolloid — Resembling a corolla; petal-like.
Cordate — Heart-shaped.
Corymb — A raceme in which the lower or outer pedicels are longer than the inner, producing a flat-topped inflorescence, the outer flowers opening first.
Crown — A corona between the petals and the anthers of some milkweeds (Asclepiadaceae).
    That portion of a stem at ground level.
    The leafy portion of a tree.

Cyme — A flat-topped inflorescence in which all the pedicels arise from the same level on the stem. The central flowers open first.

Deciduous — Falling, not remaining attached. Said of petals or sepals that drop soon after opening or of leaves that fall in the fall.

Decumbent — Reclining, as a stem that leans against other vegetation but with an upright or rising tip.

Dehiscent — Splitting, as a dry fruit, releasing the seed or seeds.

Edible — Can be eaten. This doesn't mean that it should be eaten or that it will taste good if you do eat it! You are on your own!

Drupe — A fleshy fruit containing a single stone, or in a few cases, with several free 1-seeded stones.

Entire — Without indentation, said of smooth leaf margins.

Ephemeral — Of short duration.

Erect — Essentially vertical.

Filament — The anther-bearing stalk of a stamen.

Follicle — A dry fruit consisting of a single carpel, generally with several to many seeds, and opening by a slit.

Head — An inflorescence consisting of several to many flowers on a common receptacle and surrounded by one or more series of bracts. Generally used also to denote any cluster of closely placed flowers, not necessarily having bracts or a common receptacle.

Herbaceous — Like a herb, not shrubby or tree-like.

Indehiscent — Said of a fruit that does not open at maturity.

Inferior — Said of an ovary that is below the attachment of the calyx and corolla.

Inflorescence — The flowering part of a plant.

Involucre — A row of bracts, either separate or united, surrounding the base of a flower or group of flowers.

Irregular — Not all alike.

Legume — A one-celled, generally several-seeded, fruit with two equal halves (each half known as a valve), splitting longitudinally along an upper line and a lower line between the valves when mature.

Lip — The upper or lower portion of an irregular calyx or corolla.
Lobe — An unbroken projection of the margin of united calyx or corolla, or of the margin of a leaf. The differentiation between a lobe and a tooth is not always clear.
-merous — Having parts, such as 5-merous (having 5 parts).
Obcordate — Heart shaped with the broad end apical and the narrow end basal.
Opposite — Two structures, such as leaves, arising from opposite sides of the same node.
Ovary — The seed-bearing part of a pistil.
Ovule — An outgrowth of the ovary of a seed plant that encloses an embryo sac within a nucellus.
Palmate — Said of a leaf in which 3 or more leaflets radiate from a common point.
Panicle — A branched, or compound, raceme.
Pedicel — The stalk of a single flower.
Pericarp — Refers to a fruit; the ripened walls of an ovary.
Persistent — Not deciduous. Remaining attached as the calyx of some fruits.
Petal — One of the members of a corolla of a flower, interpreted variously as a modified leaf or a modified stamen.
Petaloid — Resembling petals, as the colored calyx of some flowers.
Petiolate — Having a petiole.
Petiole — The stalk or stem of a leaf.
Pinnate — A compound leaf with leaflets along each side of a central axis.
Odd-pinnate, with a terminal leaflet; even-pinnate, without a terminal leaflet.
Pinnatifid — Said of a leaf with the margin deeply incised or lobed in a pinnate manner.
Pistil — The ovule-bearing organ of a seed plant consisting of the ovary with its appendages.
Pod — A dry, dehiscent fruit, elongated or not.
Pollinia — The paired pollen sacs of the Asclepiadaceae.
Pubescent — Having short, soft hairs.

Raceme — A simple inflorescence of pedicillate flowers along a central elongated axis. Like a spike except the individual flowers have pedicels.
Receptacle — The expanded portion of a pedicel bearing the organs of a flower, or of numerous florets as in the Compositae.
Reflexed — Bent sharply back and downward.
Regular — Said of a flower in which all the parts of each series are similar.
Rotate — Said of a flower with united petals in which the lobes of the corolla spread maximally, like a wheel.
Sagittate — Shaped like an arrowhead.
Scabrous — Rough to the touch.
Scorpioid — Said of a coiled inflorescence, the central axis uncoiling as the flowers mature.
Sepal — One of the modified leaves composing a calyx.
Septum — A partition within an organ.
Sessile — Without a stalk; that is, without petiole or pedicel.
Silique — A many-seeded capsule with central septum. The two outer covers (valves) open from the bottom to the top releasing the seeds and leaving the septum in place.
Simple — Any plant organ, such as a stem, leaf, pistil, etc. that is not compound.
Spike — An inflorescence in which the individual flowers are sessile on an elongated central axis.
Spur — A tube-like extension of a petal or sepal of certain flowers.
Stamen — The organ of a seed plant that produces the male gamete consisting of an anther and a filament and is morphologically a spore-bearing leaf.
Stigma — The part of a flower that receives the pollen grains and on which they germinate.
Stipule — An appendage at the base of a leaf petiole.
Style — A filiform elongation of a plant ovary bearing a stigma at its apex.
Subtend — Below and close to.
Superior — An ovary that is above the attachment point of the calyx and corolla.

Tendril — A slender twining or coiling organ on the stem of some climbing plants. A modified leaf.
Terminal — At the apex.
Trifoliolate — Said of a compound leaf with three leaflets.
Tubular — Elongated and hollow.
Umbel — An inflorescence in which the pedicels of the flowers originate from a common level in such manner as to form a flat-topped inflorescence.
Unisexual — Having either only male or female parts.
Whorled — Three or more leaves originating at a common level around a stem.

Flowering Schedule

The flowering schedules are adjusted for the central region of the Llano Estacado. Flowering occurs earlier than depicted in the south and later in the north. The schedules are approximations but they should be useful to the reader in ascertaining when a plant is not likely to be in flower. The "x" indicates high flowering activity and the "o" and "." indicate progressively less activity.

Spring Flowers

```
                             J  F  M  A  M  J  J  A  S  O  N  D
Sisymbrium
    TANSEY MUSTARD           .. ooo oxx ooo ... .
Erodium
    PIN CLOVER               .. ...  ..o xoo ... .
Taraxacum
    DANDELION                .  ...  ..o xxx ooo ... ... ..
Oxalis
    WOOD SORRELL             ..  .ox xoo ... .
Lithospermum
    APACHE TEA               .   .ox xxx ..
Lamium
    HENBIT                       ..o xo.
Lesquerella
    BLADDER POD                  ..o xoo ..
Linum pratense
    BLUE FLAX                     .  xxx oo. .
Verbena
    PRAIRIE VERBENA               .  oxx xxx xxx o.. ...
Allium
    WILD ONION                    .  oxx ... .
Astragalus
    LOCO WEED                     .  xxx xxo .
Corydalis
    SCRAMBLED EGGS                .  xoo oo.
Berberis
    AGARITA                       .  xx.
```

Summer Flowers

	J	F	M	A	M	J	J	A	S	O	N	D

Castilleja
    INDIAN PAINTBRUSH .xx o..

Penstemon
    BEARD-TONGUE .ox xx.

Machaeranthera
    TANSEY ASTER ..x xxo o.. ... .

Melampodium
    MOUNTAIN DAISY .ox xxo o.. ... .

Quincula lobata
    PRAIRIE LANTERN ..x xxx xxo ooo ooo o.

Dalea frutescens
    DALEA .xo oo. ... ...

Engelmannia
    CUT-LEAF DAISY .x xxo ooo ooo ... ..

Chamaesaracha
    FALSE NIGHTSHADE .o xxx ooo ooo ... ..

Haplopappus
    HAPLOPAPPUS .x xxx xo. ... .

Gaura
    GAURA .x xxo ooo ooo ooo oo

Berlandiera
    GREEN EYES .x xxo oo. ... .

Senecio
    THREADLEAF GROUNDSEL .x xxo ... . ..o xo

Sphaeralcea
    COPPER MALLOW .x xo ... .

Sisyrinchium
    BLUE-EYED GRASS .x xo.

Gaillardia
    FIRE WHEEL .x xo. ...

Nothoscordium
    CROW POISON .x o.

Polygala
    MILKWORT .x xxo ..

Teucrum
    GERMANDER . xxx ..

	J F M A M J J A S O N D
Linum Rigidum **YELLOW FLAX**	. xoo ... ...
Cirsium **TEXAS THISTLE**	. xxo ...
Argemone **PRICKLY POPPY**	. oxx xxx xoo ... .
Ratibida **MEXCIAN HAT**	. oxx xo. ... .
Dithyrea **SPECTACLE POD**	. oxx xxx xxo o.. .
Tradescantia **SPIDERWORT**	. ox. ..
Delphinium **PRAIRIE LARKSPUR**	. xx. .
Echinocactus **DEVIL'S HEAD**	. xo.
Echinocereus **HEDGEHOG CATCUS**	. xo.
Nama **TINY 'TUNIA**	. xxx xoo oo. ...
Hymenopappus **OLD PLAINSMAN**	. xxo o.
Zinnia **ZINNIA**	. xxx oo. ... .
Sophora **WHITE LOCO**	. xo.
Melilotus **SWEET CLOVER**	. xxx xxo oo. ... ..
Gilia **BLUE GILIA**	. xo. ... ...
Psilostrophe **PAPER DAISY**	. xxx x.. ...
Thelesperma **RAYLESS**	oxx oo. ... ...
Schrankia **SENSITIVE BRIER**	.xx xxx o.. ... ...

	J	F	M	A	M	J	J	A	S	O	N	D

Asclepias
GREEN MILKWEED .oo xxx xxx ooo .

Solanum
TROMPILLO oxx xxx xxx xoo o.

Hoffmanseggia
RUSH PEA oxx xxx x.. ... ..

Monarda
HORSEMINT oxx o.

Medicago
ALFALFA .xx xxx xx. ... ..

Lygodesmia
SKELETON PLANT .ox xoo ... ...

Helianthus Annuus
SUNFLOWER .ox xxx xxx ooo o..

Yucca
SPANISH BAYONET .xx .

Centaurea
STAR THISTLE ..x xoo ..

Xanthisma
SLEEPY DAISY .xx ooo ooo ...

Mentzelia
STICKLEAF .ox xxo oo. ... ..

Opuntia
PRICKLY PEAR .xx x.

Opuntia
CHOLLA .xx xo.

Cucurbita
BUFFALO GOURD .ox xxo ..

Helianthus ciliaris
BLUE WEED .o xoo ... ...

Commelina
DAY FLOWER . xoo ooo ooo .

Tribulus
GOAT HEAD . xxx xxx xxo ..

Fall Flowers

	J	F	M	A	M	J	J	A	S	O	N	D
Polygonum **KNOT WEED**							...	..x	xxx	oo		
Lythrum **LOOSESTRIFE**							...	xxo	o..	..		
Solanum **BUFFALO BUR**						.ox	xxx	xxx	oo.			
Stephanomeria **STEPHANOMERIA**						.xo	ooo	ooo	...			
Prionopsis **SAW LEAF DAISY**						.o	xxx	xxx	oo			
Proposcidea **DEVIL'S CLAW**						o	xxx	xxx	o.			
Vernonia **IRONWEED**							.x.	...	.			
Grindelia **CURLY-CUP GUM WEED**							..x	xx.				
Liatris **GAY FEATHER**							o	ox.	.			
Flaveria **FLAVERIA**							.	xxx	.			
Gutierrezia **BROOMWEED**								.xx	..			
Kuhnia **FALSE BONESET**								.o	xx.			

INDEX

Abronia, 35
Acacia, 41
  seyal, 42
  texensis, 41, 42
Achillea millefolium, 76
Acleisanthes longiflora, 36
Agarita, 38, C-4
Alfalfa, 44, C-7
Algarita, 38
Alismataceae, 17, **29**
Alizarin, 63
Allium, 30
Allionia incarnata, 36
Amaryllidaceae, 18, **32**
Amaryllis, 32
Ambrosia, 70
  grayi, 70
Ammannia auriculata, 51, C-11
Angel's hair, 54
Angel's trumpet, 36, C-3
Angiosperm, 13
Annual, 12
Annual aster, 15, 69, C-19
Anther, 13
Apache tea, 56, C-13
Aplopappus, 68
Argemone sclerosa, 38
Aristolochia coryi, 33
Aristolochiaceae, 20, **33**
Arrowhead plant, 8, 29, C-1
Asclepiadaceae, 24, **53**
Asclepias, 53
  engelmannia, 53, C-12
  latifolia, 53
  oenotheroides, 53
Asparagus, 30
Asteraceae, 18, **65**
Aster subulatus, 8, 15, 69
Astragalus mollissimus, 44
Balsam gourd, 64, C-17
Barberry, 38
Bastard toadflax, 33
Beard-tongue, 61, C-15
Berberidaceae, 20, **38**
Berberis trifoliata, 38

Berlandier, Jean Louis, 70
Berlandiera lyrata, 69
Bindvine, 54, C-12
Birthwort, 33
Bitterweed, 74, C-21
Bladder pod, 9, 40, C-5
Bluebell, 65
Blue curls, 55, C-13
Blue-eyed grass, 32, C-2
Blue flax, 27, 46, C-8
Blue gilia, 54, C-12
Blue weed, 72, C-20
Bluebell, 65
Borage, 24, 55
Boraginaceae, 24, **55**
Bougainvillea, 35
Bridwell formation, 1, 2, 3
Broomweed, 9, 68, C-18
Buckwheat, 34, C-2
Buffalo bur, 47, 60, C-15
Buffalo gourd, 7, 63, C-16
Bull nettle, 47, C-9
Cactaceae, 18, 50
Cactus, 50
Caliche, 9
Callirhoë involucrata, 48
Caltrop, 46
Calylophus serrulata, 52
Calyx, 13
Campanulaceae, 23, **65**
Camphor weed, 66, C-17
Canadian River, 2
Caper, 41
Capparidaceae, 21, **41**
Caprock, 1
Capsella bursa-pastoris, 40
Cardinal flower, 65, C-17
Carpel, 13
Cascabel, 77
Castilleja sessiliflora, 62
Catclaw, 41, C-5
Cenozoic, 1
Centaurea americana, 76
Chamaesaracha coronopus, 60
Cheese weed, 49, C-10

Chenopodiaceae, 18, **34**
Cholla, 50, C-10
Christmas cactus, 50
Cirsium texanum, 77
Cissus, 64
Citrullus lanatus, 64
Cladophyll, 50
Clammy plant, 41, C-5
Class, 15
Classification, 15
Clematis drummondii, 38
Cluster flaveria, 75, C-22
Cnidoscolus texanus, 47
Commandra pallida, 33
Commelina, 30
Commelinaceae, 18, **29**
Common mallow, 48, C-9
Compositae, 14, **65**
Convolvulaceae, 24, **53**
Convolvulus, 54
  arvensis, 54
  equitans, 54
Cooperia drummondii, 32
Copper mallow, 49
Coreopsis tinctoria, 72, C-20
Cornstarch, 28
Corolla, 13
Corydalis curvisliqua, 39
Cotton, 48
Couch formation, 1
Cowboy rose, 48
Cowpen daisy, 72, C-20
Crape myrtle, 51
Crazy weed, 44, C-7
Cretaceous Period, 1
Crow poison, 31, C-1
Crowfoot, 37
Cruciferae, 21, 22, **39**
Cucurbita foetidissima, 63
Curcurbitaceae, 23, **63**
Curly-cup gum weed, 67, C-18
Cuscuta, 54
Cut-leaf daisy, 70, C-19
Dalea, 43
  aurea, 43
  frutescens, 43
  jamesii, 43, C-7
Dandelion, 79, C-24

Datura wrightii, 61
Dayflower, 30
Delphinium virescens, 38
Descurainia, 39
Desert holly, 77, C-23
Devil's claw, 7, 62, C-16
Devil's gut, 54
Devil's head, 50, C-10
Dichotomous key, 17
Dicot, 33
Dicotyledoneae, 17, 18
Disc flower, 14
Dithyrea wislizeni, 40
Dodder, 53, 54, C-12
Dog weed, 74, C-22
Dutchman's pipe, 33, C-2
Dyssodia acerosa, 74
Echinocactus texensis, 50
Echinocereus reichenbachii, 51
Engelmann, George, 70
Engelmannia pinnatifida, 70
Eriogonum, 34
Erodium, 11
  cicutarium, 45
  texanum, 45
Erysimum capitatum, 40
Eupatorium, 66
Euphorbia marginata, 47
Euphorbiaceae, 19, **47**
Evening primrose, 52, C-11
Fabaceae, 22, 23, **41**
False boneset, 66, C-17
False nightshade, 60, C-15
Feather dalea, 43, C-6
Figwort, 61
Filament, 13
Firewheel, 73, C-21
Flame flower, 37, C-4
Flaveria trinervia, 75
Flax, 45
Flutter mill, 52
Figwort, 61
Four-o'clock, 35, C-3
Frog fruit, 57, C-14
Fumariaceae, 21, 22, **39**
Fumitory, 39
Gaillardia, 73
  pinnatifida, 73

pulchella, 73
Gaura, 52, C-11
  parviflora, 53
  villosa, 52
Gay feather, 9, 66, C-17
Genus, 15
Geraniaceae, 21, 22, **45**
Geranium, 11, 45
Germander, 58, C-14
Gilia rigidula, 54
Globe berry, 64
Globe mallow, 49, C-9
Goat's beard, 78, C-24
Goat head, 46, C-8
Golden dalea, 43, C-7
Goosefoot, 34
Gourd, 63
Grassland, 7
Great Plains Province, 4, 7
Green eyes, 69, C-19
Green milkweed, 53, C-12
Grindelia squarrosa, 67
Ground cherry, 59, C-14
Gutierrezia sarothrae, 68
Habitats, 7
Hackberry, 8
Haplopappus, 9, 68, C-18
  spinulosus, 68, 73
Hedgehog cactus, 51, C-11
Hedyotis, 63
  nigricans, 63
  acerosa, 63
Heel bur, 44
Helenium badium, 74
Helianthus, 71
  annuus, 71
  ciliaris, 72
  maximiliani, 71
Heliotrope, 56, C-13
Heliotropium, 56
Henbit, 58, C-14
Heterotheca, 66
Hibiscus, 48
Hierba de la hormiga, 36, C-3
Hoffmanseggia glauca, 42
Horse crippler, 50
Horsemint, 58
Hydrophyllaceae, 25, **55**

Hymenopappus, 75
  flavescens, 75, 76
  tenuifolius, 75, 76
Hymenoxys, 74
Ibervillea lindheimeri, 64
Ice plant, 72
Indian paintbrush, 62, C-16
Inflorescence, 14
Iridaceae, 18, 32
Iris, 32
Ironweed, 65, C-17
Isocoma wrightii, 68, C-18
Jimson weed, 61, C-15
Kansan Biotic Province, 7
Kingdom, 15
Knotweed, 8, 34, C-2
Krameriaceae, 20, **44**
Krameria, 44
  argentea, 45
  lanceolata, 44
  triandra, 45
Kuhnia, 66
Labiatae, **57**
Lactuca, 78
  sativa, 78
Lamiaceae, 23, **57**
Lamium amplexicaule, 58
Legume, 41
Leguminosae, **41**
Lemon beebalm, 58, C-14
Lesquerella, 40
Lettuce, 78, 79, C-24
Liatris punctata, 66
Liliaceae, 18, 30
Lily, 30
Linaceae, 21, **45**
Linum, 46
  rigidum, 46
  pratense, 46
Lippia, 57
Lithospermum incisum, 56
Lizard tail, 53, C-12
Llano Estacado, 1, 2, 3, 4, 5, 7, 8, 9, 11, 15, 25
Loasaceae, 20, **49**
Lobelia cardinalis, 65
Loco weed, 44, C-7
Long-day plant, 11

Loosestrife, 51, C-11
Lygodesmia, 78
  pauciflora, 78
  texana, 78
Lythraceae, 21, **51**
Lythrum dacotanum, 51
Machaeranthera tanacetifolia, 68
Madder, 63
Mala mujer, 47
Mallow, 48
Malva neglecta, 48
Malvaceae, 23, **48**
Martyniaceae, 22, **62**
Medicago sativa, 44
Melampodium cinereum, 69
Melilotus, 42
  albus, 43
  officinalis, 42, 43
Mentzelia, 49
Mexican hat, 71, C-20
Miocene, 1
Milfoil, 76
Milkweed, 53
Milkwort, 47, C-9
Mint, 57
Miriabilis albida, 35
Monarda citriodora, 58
Monocot, 29
Monocotyledoneae, 17
Morning glory, 13, 53
Moss rose, 37
Mountain daisy, 69, C-19
Muleshoe National Wildlife
  Refuge, 7
Mustard, 39
Nama hispidum, 55
Neutral plant, 11
Nightshade, 11, 58
Nothoscordum bivalve, 31
Nyctaginaceae, 19, **35**
Oenothera, 52
  canescens, 52
  jamesii, 52
  missouriensis, 52
Ogallala, 1
Okra, 62
Old man's beard, 38
Old plainsman, 75, C-22

Old red eye, 73, C-21
Onagraceae, 20, **52**
Onion, 30, C-1
Opuntia, 50
  imbricata, 50
  leptocaulis, 50
  macrorhiza, 50
Ovary, 13
Oxalidaceae, 24, **45**
Oxalis, 45
Oxytropis lambertii, 44
Palafoxia rosea, 75
Palo Duro canyon, 1, 9
Papaveraceae, 20, **38**
Paper flower, 75, C-22
Pastures, 7
Pecos River, 2
Penstemon, 61
Pentstemon, 61
Perennial, 12
Perezia nana, 77
Perianth, 13
Petal, 13
Petalostemum multiflorum, 43
Phacelia congesta, 55
Phlox, 54
Phylum, 15
Phyla incisa, 57
Physalis, 59
  viscosa, 59
Phytolaccaceae, 19, **36**
Piemelon, 64, C-16
Pigeon berry, 36, C-3
Pin clover, 45, C-8
Pistil, 13
Plains prickly pear, 50, C-10
Playa, 8
Pleistocene Epoch, 2, 3
Pliocene, 1
Plum, 8
Pokeweed, 36
Polanisia, 41
Polemoniaceae, 24, **54**
Pollen, 13
Polygala alba, 47
Polygalaceae, 23, **47**
Polygonaceae, 19, **34**
Polygonum bicornis, 34

Poppy, 38
Porcupine eggs, 61
Portulaca mundula, 37
Portulacaceae, 21, **37**
Potato, 54, 58
Prairie clover, 43, C-7
Prairie lantern, 59, C-14
Prairie larkspur, 38, C-4
Prairie sunflower, 71, C-20
Prairie verbena, 56, C-13
Precipitation, 5
Prickly poppy, 38, C-4
Prionopsis ciliata, 67
Proboscidea louisianica, 62
Psilostrophe villosa, 75
Puccoon, 56
Purslane, 37, C-3
Quinculla lobata, 59
Ragweed, 70, C-19
Rainlily, 32, C-2
Randall clay, 3
Ranunculaceae, 19, 21, 22, **37**
Rhatany, 44, C-8
Ratibida columnaris, 71
Ray flowers, 14
Rayless thelesperma, 73, C-21
Ricinus communis, 47
Rivina humilis, 36
Rocea, 68
Rock daisy, 69
Rubiaceae, 23, **63**
Rush pea, 42, C-6
Russian thistle, 35, C-2
Sage, 8
Sagittaria longiloba, 29
Salsola kali, 35
Sandhills, 8
Sand verbena, 35, C-3
Sandalwood, 33
Sand palafox, 75, C-22
Santalaceae, 19, **33**
Saw-leaf daisy, 67, C-18
Schrankia, 42
Scrambled eggs, 39, C-4
Scrophulariaceae, 23, **61**
Seasonal Succession, 11
Senecio, 76
   douglassi, 76

   riddellii, 76
Sensitive briar, 42, C-6
Sepal, 13
Shamrock, 43
Shaw, Henry, 70
Shepherd's purse, 40
Shin oak, 8
Short-day plant, 11
Sida lepidota, 49
Silver-leaf nightshade, 59, C-15
Sisymbrium, 39
Sisyrinchium, 32
Skeleton plant, 77, 78, C-23
Sleepy daisy, 67, C-18
Snake weed, 68
Sneezeweed, 74, C-21
Snow on the mountain, 47, C-9
Soap berry, 8
SOB cactus, 50
Soils, 3
Solanaceae, 25, **58**
Solanum, 11, 60
   dimidiatum, 60
   eleagnifolium, 59, 60
   rostratum, 47, 60
   triquetrum, 60
Sonchus, 78
Sophora nuttalliana, 42
Southern High Plains, vii
Sow thistle, 78, 79
Spanish bayonet, 31, C-1
Species, 15
Specific epithet, 15, 25
Spectacle pod, 40, C-5
Sphaeralcea, 49
Spiderwort, 29, 30, C-1
Sporophyte, 13
Spurge, 47
Stamen, 13
Star thistle, 76, C-23
Star violet, 63, C-16
Stephanomeria, 77
Stick-leaf, 49, C-10
Stigma, 13
Stink gourd, 63
Stork's bill, 45
Style, 13
Sumac, 8

Sunflower, 14, 65
Sweet clover, 42, C-6
Tahoka daisy, 68
Talinum lineare, 37
Tansey aster, 68, C-19
Tansey mustard, 39, C-4
Taraxacum officinale, 79
Tasajillo, 50, C-10
Temperature, 5
Teucrium laciniatum, 58
Texas thistle, 77, C-23
Thelesperma, 72, C-21
   filifolium, 72
   megapotamicum, 73
Threadleaf groundsel, 76, C-23
Tiny 'tunia, 55, C-13
Tobacco, 58, 59
Tradescantia, 30
Tragopogon dubius, 78
Triassic, 68
Tribulus terrestris, 46
Trifolium repens, 43
Trompillo, 59
Troubler of the earth, 46
Tumbleweed, 18, 35
Unicorn plant, 62
Verbena, 56, 57, C-13
   bipinnatifida, 56
   brachiata, 57
   pumila, 57
   neomexicana, 57
   plicata, 57
Verbenaceae, 24, **56**
Verbesina encelioides, 72
Vernonia, 65
Vervain, 56, 57
Wall flower, 40, C-5
Water plantain, 29
Waterleaf, 55
Watermelon, 64
Weather, 5
White ball acacia, 42, C-5
White clover, 43, C-6
White loco, 42, C-6
White nightshade, 60, C-15
Widow's tears, 30, C-1
Wine cup, 48, C-9
Wood sorrel, 45, C-8
Xanthisma texanum, 67
Yarrow, 76, C-22
Yellow flax, 46, C-8
Yucca, 31
   angustifolia, 31
   treculeana, 31
Zinnia grandiflora, 71, C-20
Zygophyllaceae, 22, **46**

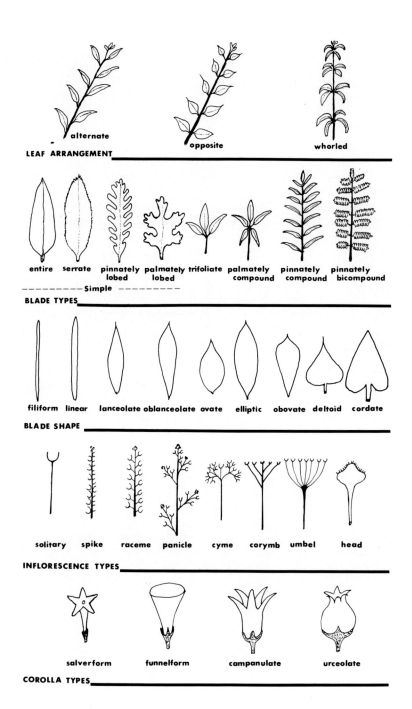

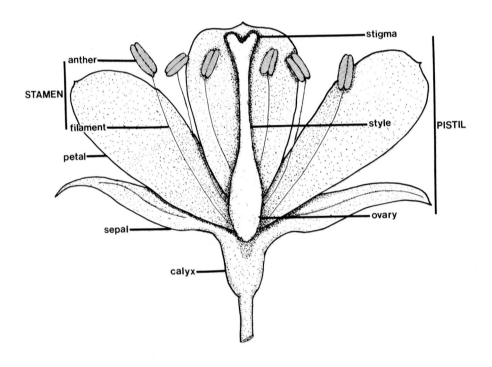